EEK

EEK!! run away!
run away!
its the cowabunga group!

D0520354

Lingua Franca Fiasco

It was my first safari in 1974, and I was eager to learn KiSwahili, the lingua franca of East Africa.

Jambo means HELLO, *Asante* denotes THANK YOU, and *Kwaheri* is GOODBYE in this beautiful, melodic language.

I would utter *Hakuna matata*, NO PROBLEM, when we'd have a flat tire, and each evening wish everyone *Lala salama*, PEACEFUL SLEEP.

As we departed one of the camps, I waved to the staff and exclaimed, *Kuhara!* Instead of waving in return, they broke into laughter. Confused, I asked what was so funny about saying, "Goodbye."

"Oh, Cowabunga," they said. "*Kwaheri* means GOODBYE. *Kuhara* means DIARRHEA."

At home in Africa.

Photo by Gary H. Lee

COWABUNGA™
SAFARIS
Africa under a rainbow

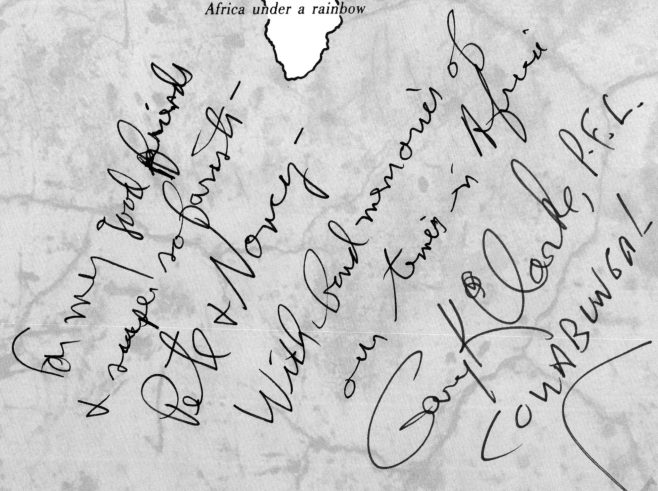

For my good friends & partners Pete & Nancy —

With fond memories of our time in Africa

Gary Clarke, P.F.C.
COWABUNGA!

This original bed sheet art graced my tent by the Zambezi River in 1994. It is my wish to be cremated in it, and my ashes returned to the campsite.

Artwork by Sarah Lamb ¦ Photo by Rod Furgason

Gary K. Clarke

President-for-Life — Cowabunga Safaris

They
Call
Me

COWABUNGA!

Once Upon A Time In Africa

Cowabunga Safaris (Pty.) Ltd.
Topeka   Nairobi   Johannesburg
2013

〆 〆 〆

Concept and Design: Gary K. Clarke
Digital Imaging: Rod Furgason
Layout & Design Coordinator: Mary Napier, Napier Communications, Inc.
Manuscript Translation: Becca Wells, C.I.H.E.
Senior Editor: Kay Quinn
Editorial Advisor: Randy Austin

〆 〆 〆

First Edition
1 3 5 7 9 8 6 4 2
Printed and bound in the United States of America by
Jostens Commercial Printing, USA

Reader Friendly

This book is meant to be an easy read. Most chapters are short, each is a complete story, and all are from the heart.

There is no sequence, so the book fits whatever situation at anytime: morning tea, favorite chair, break time, waiting room, in flight . . . even on the loo. Laughing out loud is encouraged.

੫ ੫ ੫

Photographs

The pictures in this book were chosen *not* for their photographic merit, but because they reflect the milieu of the Africa I knew and loved at the time.

I consider these photographs an expression of my feelings and experiences with the ambience and spirit of the bush, the wildlife, and the culture.

At the time, I took them more for myself than for publication, and they encapsulate some of my fondest memories of Africa. I am pleased to share them here with you.

Books by Gary K. Clarke

꘡ ꘡ ꘡

I'd Rather Be On Safari
Hey Mister—Your Alligator's Loose!
Gary Clarke's AFRICA: Wildlife, Rainbows and Laughter
They Call Me COWABUNGA! *Once Upon a Time in Africa*

꘡ ꘡ ꘡

Forthcoming

Somewhere in Africa: *serendipitous journeys of self-discovery*

To those . . .

—who have laughed and cried on safari,
who have been awed and inspired on safari.

—who have endured wind and sun, rain and
mud, dust and dung on safari.

—who have marked their territory in the bush
while listening to the night sounds of Africa.

—who have shared campfires, camaraderie
and sundowners on safari.

—who have tolerated bad puns, corny jokes
and witless witticisms on safari.

You're the Best! Cheers

True Tales from Africa

Lingua Franca Fiasco

〢 〢 〢

MISADVENTURES OUT OF AFRICA

THE BUSH LINGERS

〢 〢 〢

*I find photographing lions an easier matter
than writing books.*

—C. G. Schillings

Wildlife Photographer, Author, 1905

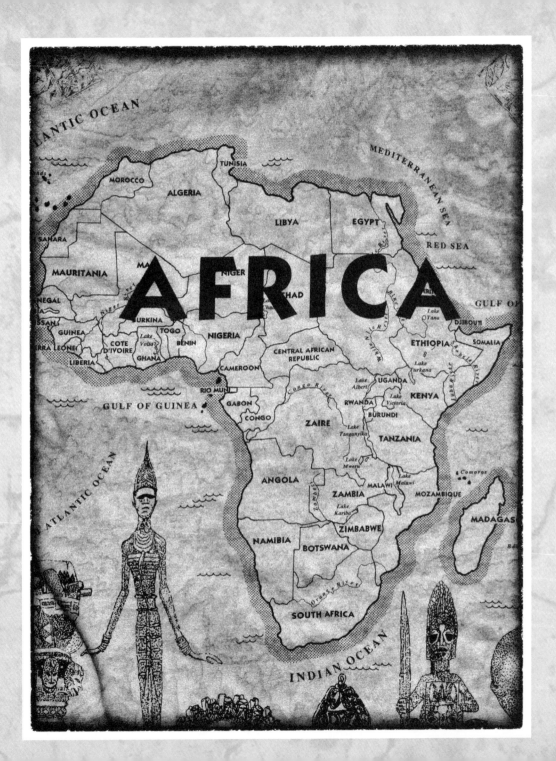

aka

In East Africa, I'm known as *Mzee Shetani*

In the Zambezi Valley, I'm the "Old Dagga Boy"

Along the Congo River, *Padre Mundele*, C.M.F

In Maasailand, *Ole Isho*

Among the Zulu, *Magangane*

At the confluence of the Blue and White Nile, *Ja'Ali*

In West Africa, "Mr. Flag"

But <u>wherever</u> in Africa . . . they call me COWABUNGA!

Who, What _is_ Cowabunga!?

What do the following have in common?

> . . . a superstitious lava flow

> . . . an aged Cape buffalo

> . . . an unlikely missionary

> . . . an unexpected honor

> . . . a mischievous personality

> . . . a courageous individual

> . . . a celebrated sundowner

After a fashion they all relate to a few of the various names bestowed upon me by different people throughout Africa.

And, of course, they call me COWABUNGA!

The Africans have such fun and games with my assorted nicknames that **who** I am, depends on **where** I am, and **when** I'm there, and even **what** local language is spoken in that part of Africa.

Who I am *not*, however, is **Mister** Clarke—he was my father. Please call me Gary.

That is my standard line in the USA, and it works with most Americans, even children.

But Africa is a different story. Initially, of all things I was addressed as **Dr.** Clarke. HA! I told them I was *not* a Doctor; I just looked like I needed one most of the time.

I think this came about because on my first safari at age 35, I was already bald-headed, with a full beard and a bit of a pot belly. The Africans assumed that I was—get this—a university professor!

Of course, that was before they had witnessed my eccentric behavior and unorthodox sense of humor. Then, their initial decorum and reserve quickly dissolved into acceptance, trust and friendship, resulting in the first of a succession of endearing nicknames.

Chief among these, in KiSwahili-speaking countries, would be "Mzee Shetani," which translates to "The Old Devil Man." How in the world did this come about? Particularly, I might add, since it is so undeserved.

My first safari was to East Africa, where the lingua franca is KiSwahili. We were in southern Kenya, at the Shetani Lava Flow, a vast expanse of black volcanic rock that erupted some four hundred years ago. The word Shetani means 'devil' in KiSwahili, and with the eruption, the locals in the area thought the devil himself had come up from the bowels of the earth—hence the name.

We stopped and alighted from our safari vehicles to take photographs. By now I felt quite comfortable teasing our drivers with such comments as, "We should have turned in the other direction," and "That's not a zebra, it's a giraffe!" I couldn't resist a few more mischievous pranks.

"Look out, there's a snake!" They all jumped, but there was no snake. "Uh-oh, we have a puncture!" They checked each of our tires and none were flat. When they realized I was up to my usual tricks they said, "You are just a Shetani, a devil for always playing jokes on us. You are Mzee Shetani." "Mzee" in KiSwahili is the term for a respected elder. In my case, it just means an old man, so as Mzee Shetani, I'm The Old Devil Man.

This was the beginning of a treasured rapport with the people of Africa, as well as a multitude of local nicknames accorded to me by them. And they have played many tricks on *me* over the years.

Most people in Kenya, Uganda, and Tanzania see humor in my Mzee Shetani nickname, particularly applied to an affable old white guy. They seem to feel a kinship in using it. Only now and then do I get a negative response. Once, in a remote village in the Northern Frontier District of Kenya, an old mama whispered to me: "That is not a proper name for you; you are a good man." I assured her it was just for laughs.

But the most fun is when I'm in a new area and people hear my KiSwahili name for the first time. They laugh, but sometimes ask with concern, "Do you know what that means?" With a glint in my eye, I strike a body builder pose and say, "Yes. Mzee Shetani means . . . a strong, handsome, intelligent guy!" After a puzzled hesitation, this elicits a guffaw, a handshake, and slap on the back!

Each of my nicknames has a story, and each one has a connotation—sometimes humorous, sometimes meaningful. Two of them have a special significance to me, and both were quite unexpected.

As a member of The Explorers Club, I participated in a momentous expedition in 2001 to retrace the footsteps of the early explorers in their search for the source of the Nile. There were 16 others in the group. We initially met in London at the Royal Geographical Society to review the archives and antiquarian map collection, with many original maps by the explorers themselves.

In keeping with historical routes, our adventure commenced on the island of Zanzibar, proceeded into the interior of the African mainland, and was scheduled to culminate in the Sudan at the confluence of the Blue and the White Nile in Khartoum.

While we were in a remote region of Tanzania, close to where Stanley found Livingstone, 9/11 happened. A few days later, in Uganda, the rest of the journey was cancelled. The world was in disarray, and everyone was frantic to get home . . . except me.

Despite the dire warnings, despite the fact that I knew no one, I ventured into the Sudan—alone. On my arrival in Khartoum, the first words spoken to me were: "You already have a reputation and a name, in the Sudan. You are *Ja'Ali*, 'The Courageous One.' Welcome." That was the beginning of an incredible experience.

The other unexpected name reflects the diversity of the continent. For eons, the Maasai culture of East Africa has observed four life stages for the male gender: Childhood, Initiation, Warrior rank, and finally, the status of Elder.

On 6 June 2002, in a surprise ceremony in southwest Kenya, I was inducted as a Maasai Elder of the Oltukai Mara Clan. My name and credentials had been submitted months earlier, unknown to me, and carefully reviewed by the Council of Elders from the Clan. They granted unanimous approval, "In recognition of his 100 safaris to Africa, and his contributions to Kenya and its peoples."

All of the Senior Elders were present for the heartfelt induction rite, which was solemn yet joyous. I was presented with my own traditional, authentic trappings of my new status, and bestowed with a Maasai name: *Ole Isho,* "the one who gives."

Those who know me will attest that seldom am I at a loss for words, but this honor caught me totally unawares. Fortunately, I knew just enough of the Maa language to respond properly and express my appreciation. I even elicited a laugh from the Elders when I asked, in conclusion, "Now, as a proper Elder, where are my cows . . . and my wives?"

Some of the other local names conferred on me in various regions of Africa include *Magangane*, Zulu for "the naughty one" or a mischievous individual; Mr. Flag, because I always had a Flag beer for a sundowner in West African countries; *Padre Mundele*, "The White Father," when I traveled to Zaire (or DRC) disguised as a priest; and The Old Dagga Boy. This last one is especially prevalent in Southern Africa and needs a bit more explanation.

In the Zambezi Valley countries of Zimbabwe and Zambia, as well as in Botswana and South Africa, "Old Dagga Boy" is a term used to denote a male Cape buffalo in his senior years, set in his ways, often ostracized from the herd, but *not* past his prime. Battle scarred and cantankerous, such an individual often has sparse hair and will roll in mud to cool his aged body and protect it against the fierce African sun. A local term for mud is 'dagga,' hence the name. To see one of these venerable old guys—his ravaged body caked in mud, decrepitly ambling with a broken horn, a tattered ear, a disfigured eye—is to see a statement of survival, the essence of the daily struggle of life in the African bush.

While I could never begin to claim the intrinsic nature of an Old Dagga Boy, I do share a few physical characteristics—and even some behaviors. At every opportunity, and especially on canoeing safaris, I like to wallow deep in Africa's thickest and ooziest mud—**glorious mud!**—especially if it has been churned by elephants, rhinos, warthogs, and other Old Dagga Boys. Soothing and therapeutic, it leaves me refreshed and energized.

All of these names indicate an acceptance by the Africans and each is considered an honor. Yet, while each name applies only to a certain area, there is one name that is used no matter where I am in Africa: COWABUNGA! Why so?

Among other things, perhaps it is the phonetic aspect of this multi-syllable word, as well as the fact that no one else has such a name.

It all started decades ago as a naive antic. A zoo colleague sent to me a poster depicting a pair of copulating baboons, with the word COWABUNGA at the top. It was a good "inside zoo joke," so I took copies to the next zoo directors' conference, thinking some might find it mildly amusing. To my surprise, it was an instant hit! *Everyone* wanted a poster. When delegates would see me, instead of saying "Hello," they would exclaim, "COWABUNGA!"

And it came to pass that COWABUNGA posters, printed at my own expense, were distributed to zoo people across the land—not just staff, but volunteers as well. Many told me they hung them on the inside of the bathroom door in order to beguile guests. Zoo docents across the country used it as a code word to alert associates giving school tours, something like "zebra

Cowabunga around the next corner."

I began to receive mail addressed to "Cowabunga Clarke," and people, some I'd never met, began sending me photos they had taken or pictures they came across of various species of animals breeding—both wild and domestic—for my "Cowabunga collection." Even the staid and dignified Brookfield Zoo in Chicago succumbed to the mania. On the occasion of my 25th Anniversary in the zoo profession, I received a package in a plain brown wrapper. Dubbed the "Cowabunga Chronicles," it was a series of superb black-and-white photos taken at Brookfield of various animals in COWABUNGA mode, artistically mounted in an accordion fold on decorative fabric panels. It was all marvelous, light-hearted fun.

The posters were a hit in Africa as well. Everyone wanted a copy and the word COWABUNGA quickly caught on. Camps and lodges referred to us as "the Cowabunga Group." On game drives, guides and drivers would say, "Look folks, giraffe Cowabunga!"

At the end of 1989, I retired from 'zoo biz' after 32 years in the profession, 26 of them as Zoo Director in Topeka, to start my own safari company.

What to call it?

The name would be very important from a marketing point of view, as it would represent us to potential clients we had never met. The name should reflect quality and adventure, and be distinctive among intense competition world wide—from the multitude of local and national travel agents, to the international tour operators and long established, specialized safari companies.

So, what to call it?

Sunset Safaris? No.

African Journeys? No.

Thorn Tree Safaris? No.

Wildlife Adventures? No.

Big Five Safaris? Already taken.

I spent days racking my brain with dozens of proposed names, and none seemed just right. But what was not obvious to *me,*

seemed evident to everyone else. *All* of my zoo colleagues in the USA, and *all* of my friends in Africa, told me that I dare not call my company anything other than: **Cowabunga Safaris.**

So I did.

And it worked!

Then, however, the question posed to me, by safari clients and Africans alike, was: "But what does Cowabunga mean?"

Most safarists who asked had *not* seen the poster. Some Africans had, but Cowabunga was such an unfamiliar term that they wondered, "Is that our language?" While it does have an African sound, the answer is no, it is not an African language. At this point, a history of the word may be in order.

It first appeared on "The Howdy Doody Show," a 1950's children's television program. "Cowabunga" was the greeting exchanged by the characters Buffalo Bob and Chief Thunderthud, although it was spelled "Kawabonga."

OK, so what does Cowabunga mean?

Surfers in California yell "Cowabunga" while riding the

crest of a giant ocean wave. Snoopy, the philosophical dog in the *Peanuts* comic strip, shouts "Cowabunga" on the front of a greeting card and inside he says with a smug, "That's how we cool dudes say Happy Birthday." I receive dozens of these cards annually. And the Teenage Mutant Ninja Turtles greet each other with, "Cowabunga, Dude!"

Yes, but: WHAT – DOES – COWABUNGA – MEAN?

It is used as an exclamation of pleasure or triumph, and I am proud to say it has come to denote "the very best of safaris."

Yes, but: WHAT – DOES – COWABUNGA – MEAN?

I referenced many dictionaries, including the authoritative *Oxford English Dictionary*: not there.

Ergo, I assumed the responsibility of composing a definition.

This was a daunting task, and the circumstances had to be just right to stimulate my most resourceful creativity.

One night, by the campfire, on the banks of the Zambezi River, after many sundowners, it came to me.

> **COWABUNGA**
> An ancient tradition from the origins of life
> in the heart of Africa that has since
> proliferated into an accepted custom
> in every culture on the face of the globe.

My official business card has this definition on the front, and I always ask that an individual read it carefully, as these are rather sophisticated words, especially coming from me. Some even read it twice—more slowly the second time—yet they still ask: "But what does it mean?"

I ask them to turn the card over. On the flip side is a photo of breeding rhinos! The mysterious verbiage is immediately crystal clear. And I chose rhinos for the photo because they are so horny. These cards are in great demand and highly prized, especially in Africa. It is not uncommon when, once again, I see someone I first met 25 or more years ago, to have them show me a well-worn card and proudly proclaim: "See, Cowabunga; I still have your card!"

Now, for an official logo—a distinctive design that would identify Cowabunga Safaris as an organization unlike any other.

Initially, I considered an animal as a symbol, but which one? Other safari companies had used an elephant or a lion, and I did not want to duplicate. Maybe a hippo? or a hyena? No.

Ah-ha! How about the dung beetle? They are industrious, take care of details, put up with a lot of shit, and know Africa from the ground up. What more could one want in a safari operator? I seriously considered it, but in the end, decided not to use an animal.

I wanted something symbolic of the safari experience. Others had used a sunset, or a thorn tree . . . so I chose a flat tire. My son John did the art work. I liked it, but discovered that most people did not understand it, or failed to appreciate the humor.

I ruminated on a different approach, one that would immediately say 'Africa.' AFRICA? Of course—a silhouette of the continent itself—and it would appeal to my friends *in* Africa as well as my safarists. Lynn Scannell finalized the design by superimposing COWABUNG SAFARIS and our motto, *Africa under a rainbow*, over the silhouette. It works well in black-and-white, but it is glorious in color. I carefully chose the colors—red, green, black, white—as at least one is represented in the flags of the African nations where we conduct safaris. Each time the Africans see our logo on a window decal, tee shirt, envelope, bumper sticker, or whatever, they immediately relate. And the Cowabunga Safaris banner, emblazoned with the full-color logo, is proudly displayed at camps and lodges, on vehicles and boats, even in flight on hot air balloon baskets. I am frequently enjoined by the Africans to, "leave your banner with us, Cowabunga, so it will be here for you next time."

Originally, I had planned to operate Cowabunga Safaris from

my home, but friend Randy Austin felt I needed to be in a public location. As the owner of Fairlawn Plaza Shopping Center in Topeka, he had a small rental space available. Perfect.

On 1 January 1990, we established territory in a bare room 9' by 27' with nothing more than a folding card table, a rickety chair, a telephone, and a map of Africa. With the help of Randy and many others in the community, this humble beginning has blossomed into . . . *not* "an office" . . . but COWABUNGA SAFARIS MAIN CAMP. It is dripping with Africa, every nook and cranny filled with safari photos, wildlife paintings, animal sculptures, antiquarian maps, plants, skulls, biological and cultural artifacts, and mementos from my 140 safaris, some hanging from the ceiling. It may be the only place in Topeka, or Kansas, or the USA, with a map of Africa on the *outside*.

Since Cowabunga Safaris never had any paid employees, Main Camp was operated by volunteers, known as Star Guards. They were all experienced safarists with a love of Africa, and deserve continued credit and appreciation, as they tended to daily duties while I was on safari leading groups.

Main Camp is a good place to be when you are not on safari. We have hot tea and rusks in the morning, sundowners in the evening, and campfire talk any time. And we answer the phone with, "*Jambo.*"

Emanating from Main Camp was the Voice of Cowabunga, *The JUST NOW News* (JNN). Not knowing exactly when we might be able to write, publish, and mail something to deserving recipients around the world, I felt we needed an appropriate title. "Just Now" is one of those delightful expressions used in Africa and loosely defined as, "an indefinite term meaning any time from now on." Hence, if you ask for hot tea or a cold beer, you will be assured that "it is coming just now," which could mean two minutes, or two hours, or two days, or . . . well, you get the point. The masthead contained the following assertion: "*The JUST NOW News* is a now-and-then Newsletter for alumni and friends of COWABUNGA SAFARIS published whenever we have enough news and time to put it together." It was issued quarterly. Between JNN Editions, we issued a "Safari Trails" mailing.

The JNN was written and edited by Gary K. Clarke, President-for-Life, and Nancy Cherry, Administrative Officer of Cowabunga Safaris. When Brian Hesse joined Cowabunga as a Safari Leader in 1996, he was on the JNN team as well, and actually made it happen in the latter years. The JNN benefited from the talents of many people: Phil Grecian of Harry Turner & Associates, and later Debbie Scanland; Doug and Irene Hommert of Kirkwood, MO, safari alumni who maintained our computerized mailing list and provided the labels; Blenda Blankenship, alumnae and Main Camp Star Guard, who labeled and prepared each mailing according to postal regulations; and many other unsung heroes.

The JUST NOW News was published from 1990 to 2006. A rare complete set would be quite valuable today, containing much of the history and character of Cowabunga Safaris.

In 1991, it was suggested that the Cowabunga Safaris alumni, dating back to 1974, gather for a weekly luncheon to reminisce about past safaris and dream about future journeys. Remembering that writers for *The New Yorker* used to do "The

Algonquin Round Table" in New York, I decided to call ours "The Cowabunga Square Table." To this day we meet every Wednesday noon at Pizagel's Bakery in Topeka.

It is very informal—no R.S.V.P. and no regrets. A table is reserved and we have anywhere from 6 to 14 attendees, sometimes more. Out-of-town alumni schedule cross-country trips so they will be in Topeka at noon on a Wednesday. And one couple—the Hommerts from Kirkwood, MO—a suburb of St. Louis—will rise before dawn, drive to Topeka for Square Table, and return home the same day, a round-trip of over 600 miles!

I would be remiss if I didn't mention the Cowabunga Safaris Official Vehicle, my 1985 Jeep, one of the last of the true CJ models. It is named "The Blue Dung Beetle" due to its color, and has a "Jambo" sign in the front window and a Tusker Beer tire cover on the spare. A metal Maasai cow bell and a wooden Samburu camel bell dangle from the grab bars, so I and my passengers can enjoy the sounds of Africa with every bump. You can see a photo of the Jeep in front of Main Camp on page 204 of *Gary Clarke's AFRICA*, and can read a short story about it on page 145 of my book, *I'd Rather Be On Safari.*

Thank you, dear reader, for staying with me, as there is more to Cowabunga! than one might imagine. Nevertheless, you now have the background to understand some of the 'inside terms' that may pop up in the rest of the book.

All of my African nicknames are endearing to me, but I must confess that my heart is singing when . . . they call me COWABUNGA!

൹ ൹ ൹

Between Two Worlds

It is dark outside.

<u>So</u> **black**, like outer space.

It is cold outside.

Flesh-freezing cold. *Instant death cold.*

It is dark inside as well. But I have a small flashlight projecting a thin beam of light.

It is cold inside, but not life-threatening. A blanket over my shoulders wards off the chill.

"Outside" is somewhere over the vast Atlantic Ocean, forty-one thousand feet above planet earth, with no consciousness of earth or sky. "Inside" is a long metal tube, full of people. The tube has wings and four powerful jet engines. A Boeing 747.

The only noise is the resistance of the wind as it is pierced by the sleek skin of the aircraft at over six hundred miles an hour. And of course, the "white noise" from the constant drone of the engines. The sound is not irritating—it is actually reassuring, a sense of security in dependent circumstances.

As we cruise in the dark approaching the speed of sound, it is hard to get a feeling of velocity, or even motion. I am in a state of deep stillness . . . it seems I am stationary. Only now and then, there are "bumps"—clear-air-turbulence. The Captain apologizes, and calls it "light chop." I like it when this happens, and close my eyes to concentrate on the feeling.

The chop brings a bit of life to this huge technological bird and makes it seem fallible, although inanimate. Ships have their own particular motion as they resist unstable waters; trains have their own particular motion as they sway on iron rails. So too, should aircraft have a characteristic "motion"—a kinetic response to unseen irregularities in the air.

Here in the stratosphere, I am far removed from my life in Kansas— family, home, associates. Here in the stratosphere, I am also far removed from my life in Africa: friends, wildlife, wilderness.

I am momentarily suspended between the reality of my two

worlds. Nothing in this encapsulated environment relates to either. Somehow I am thrust from yesterday into tomorrow, with no sense of time or place. *Now* does not exist.

Yet, in actuality, it does. This journey from the American East coast to South Africa will traverse nearly 8,000 miles in 14 hours and 30 minutes without interruption. To me that is almost incomprehensible. What used to take years or months to achieve can now be done in days or hours.

For nearly 30 years, I have regularly made this trans-Atlantic crossing by air—now well over two hundred times. But regardless of how many more times I may do so, I will never take it for granted.

Each time I marvel at the wonder of it all . . . that it can happen . . . that I am a part of it . . . that in some indefinite form, I have a suspended existence . . . between two worlds.

卍 卍 卍

At dusk, around a thorn-log fire . . . Africa changes you in ways you do not understand.

Serengeti/Tanzania

I never tire of the drama played out by wildlife at a waterhole. A herd of dainty springbok, a group of stately greater kudu, and a single gemsbok—pregnant—embody the fleeting and the timeless of Africa.

Etosha/Namibia

At times I feel the mood of Africa to be that of primordial Eden.

Tarangire/Tanzania

These poached rhino skulls offer mute testimony to the irreversible emptiness of extinction.

Serengeti/Tanzania

Night in Africa brings a heightened sense of awareness, and one listens in mystery and apprehension . . . listens . . .

Aberdares/Kenya

These Samburu women were relocating to a new village, and donkeys laden with their worldly posessions were just out of camera range. Surviving in this harsh environment requires resilience essential for a rigorous nomadic existence.

Northern Frontier District/Kenya

From the co-pilot seat of a small plane, I marveled at a herd of Cape buffalo crossing the Zambezi River.
The aged and high-status males are positioned at the rear of the column.

Mana Pools/Zimbabwe

The "Little Man on the Crater Rim."

Ngorongoro/Tanzania

The Lighter Side

Swahili Speaking Tee Shirts

Please, do not expect me to be the well-dressed Safari Leader. You will be disappointed.

My everyday attire for the bush is a pair of strap sandals, shorts and tee shirt. More often than not, the tee shirt will be a one-of-a-kind that I have originated, with a special message for the Africans, frequently in *their* local language—be it KiSwahili or some other dialect.

This has been an extremely effective means of connecting with people of different cultures and dealing with them in a tactful, often humorous way. I call it "Tee Shirt Diplomacy," and feel it should be an official facet of global relations.

Before I give you some examples of my unprecedented tee shirts, permit me to share a joint prank played on me by one of my groups in cahoots with some of my African friends. It was my birthday and we were at Ndutu Lodge in the southern Serengeti. After dinner, the lodge staff surprised me with a birthday cake. Then my group surprised me, thanks to safarist JoAnn Myers. Each member revealed they were wearing a limited-edition Cowabunga Safaris tee shirt featuring our original logo—a flat tire!

How clever, and it *was* a surprise.

When they presented me with my very own limited-edition tee shirt I excitedly put it on, not realizing it had a message on the back, in KiSwahili, which read:

I am their leader.

Please tell me where I am.

Immediately the lodge staff, with beaming smiles and twinkling eyes, rushed over and ceremoniously escorted me to a huge wall map of the African continent.

"Now, Cowabunga," they said, speaking and pointing together.

"*This* is Africa . . .

"*This* is Tanzania . . .

"*This* is Serengeti . . .

"*This* is Ndutu . . . .

"You are *here,* with us!"

Peels of laughter erupted, together with back-slapping, singing, and dancing—a joyous celebration because Cowabunga had been tricked by a tee shirt!

And each time I returned to Ndutu, the staff would re-enact the entire episode, to the joy of new staff members and the puzzlement of my current group—even if I was *not* wearing my limited-edition tee shirt.

Most of my tee shirts have the message on the front, and the Africans are always curious to read it, whether it is in their local language or in English. Sometimes they try to "sneak a glance" without me knowing, but I am always on to that. Usually, though, they present themselves in front of me to do a "deliberate" read. While Americans read a tee shirt to themselves, Africans read each word out loud, in a careful and unhurried manner. I puff up my chest to embellish this endeavor. Never before have they seen such shirts, especially on a *mzungu wasimu,* KiSwahili for "crazy white guy."

The following pages consist of a modest "fashion review" of some of my custom-made tee shirts for Africa. Please keep in mind that I am *not* a linguist. Hence, if the message on the tee shirt is in a foreign language, it may contain unintentional errors in my attempts at reverse translation.

A sample of each tee shirt will be shown, followed by my comments about the shirt and the message.

਼ ਼ ਼

The color of a given tee shirt can be as important as the message. This one is a vivid, lime green, with black lettering in the KiSwahili language. It is very popular, and a lot of fun in the East African countries of Kenya, Tanzania, and Uganda.

Often, I will cross my palms to hide the message, in anticipation of the predictable question, "Oh, Cowabunga; how are you?" Then, with a dramatic flair, I spread my arms to reveal the answer:

I Am As Happy

As A Hyena

Eating A Goat

Hyenas are not only scavengers, but efficient predators as well. They will make difficult kills of large, wild prey, and feed on a carcass with thick, tough hide and big, hard bones. Yet, when the opportunity presents itself, hyenas will feast on domestic goats—easy to catch, tender, juicy—a true delicacy, and the ultimate bliss.

Recently, I wore this shirt when I had dinner with my family at the Olive Garden restaurant in Topeka. The waiter a few tables away—a tall, young black man—kept craning his neck to read the front. As we left I stopped and asked him if he knew what it meant. "Oh, yes," he replied. "I am from Kenya, and your shirt is wonderful!"

⊡ ⊡ ⊡

Here's another favorite in East Africa, also in KiSwahili, in an eye-catching color—a deep, yet bright, carotene orange, with black lettering. I like to cover the front with my arms to build suspense, then reveal one line at a time, and watch the response.

The message reads:

I LIKE

My Cars Fast

My Beer Cold

My Women Hot

Each line brings a more enthusiastic cheer, and I sometimes pause . . . a little longer . . . before showing the last line. Of course, this particular shirt is more popular with men, but even the women find themselves laughing, and sometimes scolding me.

So many people want one—especially the guys in the roadside curio shops. "Cowabunga, I *must* have that shirt! What do you want in trade? Select anything in my shop. I'll trade you my goats, my cows. I'll trade you my wives. Cowabunga, I *must* have that shirt!"

While I appreciate their generous offers, it is more fun for me to keep the shirt to wear on future safaris.

೨೨ ೨೨ ೨೨

Now to the country of Botswana, where the language is Tswana. This tee shirt is khaki with black lettering, and has something meaningful for the people who work in the safari industry: guides, rangers, trackers, boatmen, dugout canoe polers, cooks, camp staff—all important members of the Cowabunga Safaris team in Africa.

During the safari season, they are away from home for long periods of time, and I know they miss their families and extended families back in their home village or town.

To let them know that I understand their situation, and that I appreciate all they do for my groups, I did my best to make a shirt that reads in Tswana:

HOME IS WHERE

THE HEART IS

This cultural sentiment not only elicits a smile or a nod, but often a handshake or a hug.

Tee shirt diplomacy at its best.

卬 卬 卬

Another shirt for Botswana, again in the Tswana language, but with a bit of humor. This one is bright red with white letters.

If you know my physical appearance, you'll understand the meaning. And if you *see* me in this shirt, it is even more obvious.

The top line reads BALD HEAD.

The bottom line reads BIG BELLY.

After reading this shirt, many Botswanans pat my belly, then rub the top of my head—with great delight.

If you can't laugh at yourself, who-can-you-at?

卍 卍 卍

Now, dear reader, this shirt takes a bit of explanation. In 2002, I made my first venture into the African nation of Angola, not with a group, which was too risky, but on my own. Being unfamiliar with the indigenous ethnic languages, I resorted to the residual European language from the colonial period still spoken throughout the country—Portuguese.

Wanting to make a good first impression, I decided on an introductory phrase, so I consulted some Portuguese grammar and vocabulary books, drafted my message, and put it on a deep red shirt with black letters—the colors of the Angolan flag.

Then it was off to Angola, to meet people of all walks of life, proudly wearing my tee shirt with the Portuguese greeting:

Hello! My name is
Mr. Happy! Do
you speak English?

Or so I thought . . . until an individual took me aside to tell me that what I *really* had on my tee shirt was:

Hello! My name is
Mrs. Healthy! Do
you speak English?

Now I understood all of the polite smiles and bemused head shaking. I asked where I might be able to correct it, and was told, "Oh, Mr. Cowabunga, leave it like that. Everyone finds you most enjoyable."

It was more fun than had it been correct to begin with.

෫ ෫ ෫

Asante   Ashe   Dzikomo
Ndalumba Tatenda Siyabonga
Ni itumezi Kae leboga Enkosi
...and Thank You

    If I had to pick a favorite, this might be it, because this tee shirt appeals to so many Africans across a broad spectrum of ethnic groups and geography. And, because it says "thank you" in so many ways, it reflects my travels as well as my appreciation for various cultures.

Reading each line left to right, we have the following languages:

KiSwahili	*Maa*	*Njonja*
Tonga	*Shona*	*Zulu/Ndebele*
Lozi	*Tswana*	*Xhosa*

This shirt is a deep green color with white letters. When Africans see me wearing it at a distance, it seems to have a mesmerizing allure that draws them to me within touching distance. Usually reserved, they are relaxed, yet remain polite. Even if we've never met, they often place a hand on my shoulder while reading the shirt.

Then, with great joy, they carefully enunciate each word aloud.

At the last line in English, ". . . and Thank You," they smile, laugh, shake my hand with vigor, and sometimes spontaneously embrace me.

How rewarding . . . .

卍 卍 卍

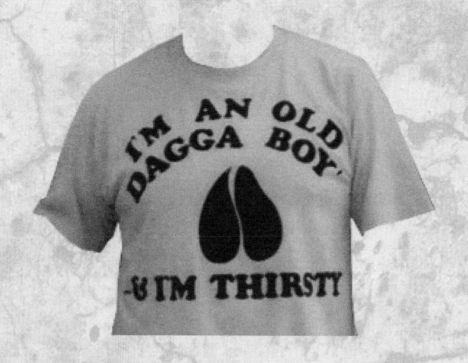

This tee shirt is khaki with black letters, and I wear it in the Zambezi Valley, where my nickname refers to an aged male Cape buffalo that wallows in the mud.

As soon as I arrive at camp, and often before I climb out of the vehicle, the staff is there with a grin—and a beer! Nothing makes me feel more welcome.

The camp staffs like to keep track of me, and they always recognize the print my sandals leave on the trails around camp. The term "spoor" is used for the tracks, paw prints, or hoof marks made by animals. When I started using a cane, the rubber tip on the end left a distinctive circle-within-a-circle imprint. One afternoon, while I was writing in my journal, I overheard two Africans talking outside my tent:

"Oh, look. There's the *new* spoor of the Old Dagga Boy."

"Yes, he went to the river."

"From there, to the lookout post."

"Ah, then to the loo."

"And now, he's in his tent."

What fun I had after that, making, with my cane, unpredictable and nonsensical spoor tracks—throughout the camp—that *surely* must have mystified them. They never said a word, but always gave me a knowing smile.

෨ ෨ ෨

I
SURVIVED
THE
GARTH THOMPSON
DEATH MARCH

Garth Thompson is a superb field naturalist, and he interprets *everything* on a walk through the bush. I call these walks "T. T. T." which stands for trees, tracks, and turds. You rise at dawn, have coffee or tea around the campfire, and you are off with Garth for a leisurely few hours in the cool of morning. No need to carry water or trail snacks, since you'll return to camp in time for a hearty breakfast.

Except . . . Garth gets so engrossed with the wonders of the bush, that he wanders further and further from camp. Time passes, and the sun gets hot, and you are thirsty and hungry and tired and your feet hurt and . . . what was to have been an enjoyable nature walk, has become a **Death March!** Finally, you get back to camp around midday, exhausted.

Granted, you do learn a lot, and on my first walk with Garth, I thought it was the exception, not the rule.

I was wrong.

So, on one of his subsequent walks, I had stashed this tee shirt in my backpack. As we neared camp on our eventual return, I lagged behind the group, and secretly donned the shirt. Just when everyone was wondering where I was, I staggered into camp, dragging my backpack, all disheveled, and wearing the tee shirt.

The group applauded me. Even Garth had to laugh.

I wish you could see this shirt in color; the red drops of blood are quite realistic.

卍 卍 卍

front

back

The guides along the Zambezi River, particularly those who conduct canoeing safaris in the Mana Pools region, are a rugged lot. They pride themselves on their ability to survive in the bush with little or no amenities. They shun things representing the easy life in towns and cities, like cold soda pop (Coke) and fresh, soft rolls (Buns). Hence, it is an insult to refer to someone as "Coke & Buns," or infer that they operate a "Coke & Buns Camp."

Knowing this, I could not resist wearing a tee shirt specifically made for my old friend and respected guide, Garth Thompson. The front read "Coke & Buns Safaris," referring to his operation, although it truly was first class. The back read "Biltong & Warm Beer Safaris," which implied that Cowabunga was tough. Biltong is like Kansas beef jerky, but often made from wild game, such as impala, eland, kudu or sable antelope.

Garth had a good laugh from my shirt, and at lunch he had the staff serve me *only* Biltong and Warm Beer—warmed with a candle! I had a good laugh at that, and I figured we were even.

I was wrong.

At dinner I was served nothing but Biltong and Warm Beer, while my group lavishly feasted on Coke & Buns of every description.

The next morning at breakfast it was the same thing for me . . . and at lunch . . . and at dinner: nothing but Biltong and Warm Beer. Of course, I gnawed on the biltong like a predator on prey, and guzzled the warm beer like a dehydrated camel. I sensed that my group was feeling a bit sorry for me, but was reluctant to enter this war of wills, which continued into the third day.

On our last night in camp, Garth relented, commended me for being so resolute and unyielding, and ordered that I be rewarded with a grand and glorious dinner. Little did he know that his faithful camp staff had been secretly feeding me all the while, behind his back.

〢 〢 〢

Some people, when they reminisce about their wonderful times on safari, look at scrapbooks, slides, digital images, videos, or re-read their safari journals. So do I.

But much of the fabric of my African memories can be found in my wardrobe of safari tee shirts.

〢 〢 〢

Leeches Make Me Crotchety

"Cowabunga! You're bleeding!"

I was in the Okavango Delta of Botswana, sitting low in a narrow dugout canoe, called a mokoro, my knees up under my chin. We had just poled into camp, and I was in such an awkward position I could not get out. A strong African lent me a hand, and as he pulled me up he repeated, "Cowabunga! You're bleeding!"

I looked at my hands, my arms and my legs, even touched my face . . . no trace of blood.

As I stepped ashore, he said in a lowered voice, "No, Cowabunga, down there," and gestured toward my crotch.

In my crotch? Ohmigod!

"Leeches," he advised me.

Instantly I recalled that classic film, *The African Queen*, and the scene when Humphrey Bogart's character confronted leeches! It is one thing to see it in the movies, and quite another to experience it personally, just here, just now—and in my crotch!

In truth, I like leeches. They are functional and fascinating creatures. But in my crotch?

I sincerely like the other creepy-crawlies of Africa as well, but not necessarily in my crotch. They are wee, but important, members of the animal kingdom, and each plays a vital role in the balance of nature. Though considered pests by some cultures, in contrast, Africans find many species, particularly certain insects, a source of nutrition. Even some of my safarists have snacked on mopane worms, among other delicacies.

My Cowabunga groups have had various encounters with the diminutive fauna of the bush. I recall one dinner by lamplight in an open-air dining room at a safari lodge in southern Kenya. It was a warm, but pleasant evening, and soup had just been served. In the darkness outside our table, we heard the buzz and hum of—what? Insects? Yes!

Suddenly the air above us was swarming with gobs of flying

insects—beetles, crickets, grasshoppers, flies, cicadas—you name it. They were not attacking us as such, but there were so many that they started dropping onto the table, and in our soup, and on the floor, everywhere! It was literally raining insects— like a biblical plague. My group was fantastic—not frightened or panicky, but amazed and amused, even feigning protection with their soup spoons.

Abruptly, it was over.

The living insects flew away, while the dead ones littered the premises. The staff brushed them off the tables, swept the floor, served soup again, and we proceeded with dinner. After such a memorable incident, we were not "bugged" again for the rest of that safari.

But that was then.

Now, however, I was experiencing my first real one-on-one with leeches—in fact, a three-on-one. On the way to camp with our mokoros in single file, we had followed a narrow water trail through the aquatic vegetation, so high that it closed out the sky. Upon entering an open lagoon, we paused to appreciate the lovely, tranquil scene, take a few photos, and a quick dip in the Okavango Delta for a swim. I can't swim, so I simply held on to the mokoro to cool off in the refreshing, crystal-clear water.

That was when it happened!

But I did not feel a thing . . . .

Why? Because the leech—a marvelous, segmented, parasitic worm with suckers at both ends—secretes an anesthetic substance that induces insensitivity to pain, as well as an anticoagulant to enable the blood to flow freely. (Historically leeches have been used in medicine for bloodletting.)

My blood was flowing freely as I followed my African friend behind a staff tent and dropped my shorts while he successfully removed the leeches—fat, black, shiny, and full of blood. I was glad to get rid of those suckers.

My shorts were so bloody that I stepped out of them, then took my shirt off and wrapped it around my waist so I could dash to my tent. At that moment Lee, the lovely camp hostess I had not yet met, came up and introduced herself (she literally caught me with my pants down . . . oh, err . . . off). She quickly understood my predicament and left to arrange for some ointment to treat my wounds.

I showered, applied the ointment, which was somewhat onerous, donned a clean pair of shorts (the second of three pairs I carried), and joined the group for lunch.

The conversation around the table included, of course, my leech episode. Several people expressed surprise that it had happened to me, commenting that I was seldom bothered by the annoying little denizens of Africa. This was true, even though I never used insect repellent (too messy). Tsetse flies are a case in point.

A large fly with long wings that lie folded over the back, tsetses feed on mammal blood (particularly warthogs) and range over savannahs, woodlands and waterways. They have a long proboscis and a stinging bite that often leaves a welt. Personally, I do *not* follow the usual advice on tsetse flies, which is:

If a tsetse fly enters your vehicle, kill it immediately;

if one lands on you, slap it once, etc.

Many times I have been in a vehicle with four or five other humans and *dozens* of tsetse flies. Everyone (except me) will be swatting the flies, which seems to make them angry. I think when you fight the tsetses, your body chemistry emits odors that agitate the flies. Now, what I am about to tell you may seem fanciful, but it is true—ask those who have been with me. I calmly sit and think peaceful thoughts. And as silly as *this* sounds, I sing a little song, very softly: "Hello, little tsetse fly; say hello, and say goodbye." While the tsetses are swarming the others, they leave me alone. If one does land on my bare skin, rather than slap it, I gently pick it up by the wings and place it outside. It works for me.

A stereotypical image of Africa is one with safarists besieged by annoying vermin of all ilk. That brings to mind a classic Gary Larson cartoon. It shows two rotund Victorian-type explorers, in pith helmets and short pants, engulfed by insects. One of them, scrutinizing the label on a can of insect repellent, is exclaiming to the other: "Wait a minute! . . . McCallister, you fool! *This* isn't what I said to bring!" The label on the can reads: ON.

Our lunch concluded, and when I stood up to leave the table, one of the ladies said, "You're bleeding again." I was, and not just a little—a lot. Even though the leeches had been removed, the anticoagulant secreted at the attachment sites resulted in continued blood flow.

So, back to my tent to clean up and put on my third (and last) pair of unsoiled shorts. The situation was now becoming awkward and embarrassing—an annoying inconvenience of the first order. But it did give me a little more insight and empathy for what women have to periodically endure in various circumstances.

Later that afternoon, our guides took the group by power boat to observe and photograph a nesting pair of African fish eagles. As I was concentrating on bringing the birds into sharp focus, one of the women said to me, "You're still bleeding."

Another woman quickly remarked, "Now, if you'd just get a headache, you'd *really* know what it is like."

と と と

Ears Pierced While You Wait

The title of this chapter is from a sign in a Botswana jewelry shop. It is what I call a T.A.B. sign. T.A.B. stands for "That's Africa, Baby."

T.A.B. would apply to the name of a business as well. One of my very favorites is *The Three Swallows Bar,* a paragon of ambiguity. If *swallows* is a verb, it means three mouthfuls of beer down the throat; if *swallows* is a noun, it means three fast-flying, forked-tailed birds.

T.A.B!

Following are more examples of bewildering T.A.B. signs.

In a restaurant in Zambia:

Open seven days a week and weekends.

On the grounds of a private school in South Africa:

No trespassing without permission.

On a window of a Nigerian shop:

Why go elsewhere to be cheated when you can come here?

On a poster in Ghana:

Are you an adult who cannot read? If so, we can help.

In a hotel in Mozambique:

Visitors are expected to complain at the office between the hours of 9:00 am and 11:00 am daily.

On a river in the Democratic Republic of Congo:

Take note: When this sign is submerged, the river is impassable.

In a Zimbabwean restaurant:

Customers who find our waitresses rude ought to see the manager.

A sign seen on a hand dryer in a Lesotho public toilet:

Risk of electric shock—Do not activate with wet hands.

On one of the buildings of a Sierra Leone hospital:

Mental Health Prevention Centre.

In a maternity ward of a clinic in Tanzania:

No children allowed!

In a cemetery in Uganda:

Persons are prohibited from picking flowers from any but their own graves.

In a Malawi hotel:

It is forbidden to steal towels, please. If you are not a person to do such a thing, please don't read this notice.

A sign posted in an Algerian tourist camping park:

It is strictly forbidden on our camping site that people of different sex, for instance a man and woman, live together in one tent unless they are married to each other for that purpose.

In a Namibian nightclub:

Ladies are not allowed to have children in the bar.

A prominent African national park (that will remain unidentified) has a sign that conveys relief, then dismay:

That some of the road conditions in South Africa are *really* bad is evidenced by this sign:

And parking can be a problem according to this sign on a four-foot high fence:

NO PARKING
ABOVE
THIS SIGN

A classic T.A.B. sign in South Africa makes you wonder if it refers to one individual or two different people. It features a Red Cross logo and reads:

DOCTOR
SURGERY +
At Prime Butchery

In Zimbabwe, at a national park along the Zambezi River, a statement in the Official Fishing Regulations Handbook advises:

*If you hook a hippo or a crocodile,
you must release it immediately.*

Sign on a remote airstrip in Botswana:

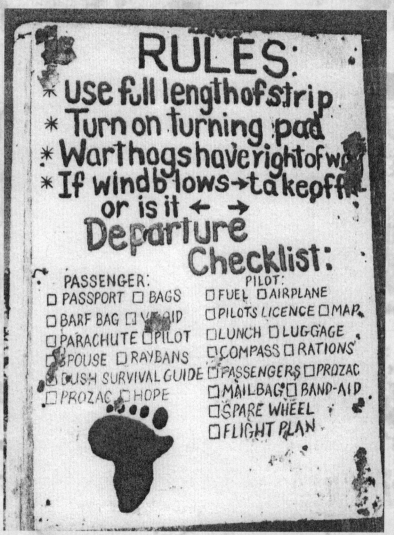

A sign at the entrance to Arusha National Park in Tanzania:

NOTICE TO ALL VISITORS:

PLEASE DO NOT

TAKE LIBERTIES WITH

WILD ANIMALS

In the Ol Pejeta Reserve of Kenya, a high, one-lane, wooden bridge spans the Uaso Nyiro River below. Posted by the approach at each end of this bridge is a sign that states:

NOTICE

THIS BRIDGE IS RATED FOR A

MAXIMUM OF 11 TONS

ELEPHANTS ARE THEREFORE

REQUESTED TO CROSS TWO

<u>AT A TIME ONLY</u>

Recently I came across an old book on Africa dating back to the late 1930s. An intriguing photograph taken in Freetown, Sierra Leone showed a nondescript wooden building with a corrugated metal roof. Extending along the roof line was a sign—in English—painted with white letters specifying the name of the business, together with some words of advice from the proprietor:

THE SYMPATHETIC UNDERTAKER
does not live like a fool and die like a big fool.

Eat and drink good stuff.

—Undertaker's Advice

T.A.B.

֏ ֏ ֏

Heebie-Jeebies with Endangered Feces

The stern U.S. Customs Agent, in crisp uniform and shiny badge, approached me, trained sniffer dog in the lead. I was at the baggage claim area of JFK International Airport in New York, having just arrived on the long flight from Johannesburg, South Africa. My carry-on bag was at my feet. The dog was dutifully sniffing luggage for illegal substances—primarily fruits, plants, and meat.

What I had in my bag was worse—much worse, so I knew I was in trouble—big trouble. My stomach churned. I was nervous and sweaty. How could it not show? Never had I suffered from such an acute case of the heebie-jeebies.

And all because of a pile of crap.

Oops! Sorry about that—let me explain.

On the last night of my safari, the bush guides along the

Zambezi River presented me with a gift. "Cowabunga," they said, as we had farewell sundowners around the campfire. "We want you to have your own midden for Main Camp." A midden is the result of a territorial marking behavior when a given species of animal repeatedly defecates in the same spot to establish a scent. In other words, a dung heap.

And this midden was not simply routine, easy-to-find dung— it was very special *dignified dung!* (Ah, such dear friends I have in Africa.)

All joking aside, I must give the guys credit. It was obvious that they had worked long and hard to collect droppings from little known and seldom seen African mammals, many of them nocturnal: aardvark, bush baby, civet, spring hare, honey badger, and porcupine, among others. Since it was rare to see droppings of these species, I called them endangered feces.

It truly was a magnificent midden and I was honored. I knew they would not do this for just anybody. What an educational dimension it would add to Cowabunga Safaris Main Camp.

Now, at JFK, all these thoughts ran through my mind as a

possible, but perhaps not plausible explanation for the midden in my luggage.

The Customs Agent stood close to me as the dog proceeded to scrutinize my bag with his highly trained nose. My heart was in my throat. I knew that the dog was trained to sit and bark when he detected the scent of a prohibited substance. I expected he would do just that at any second.

Ever so slowly, the dog made a complete circle around my bag, his keen nostrils buried in the fabric. His body was tense, and so was mine. The Customs Agent was attentive and hovering. Again, the dog made a complete circle around my bag, ever so slowly, and stopped.

Then the dog lifted one hind leg and urinated all over my bag.

Appalled, the Customs Agent exclaimed: "Oh, sir, I'm so sorry! I've never seen him do that before!"

To which I replied, "Oh, that's OK; don't worry about it."

I picked up my bag and made a grateful exit.

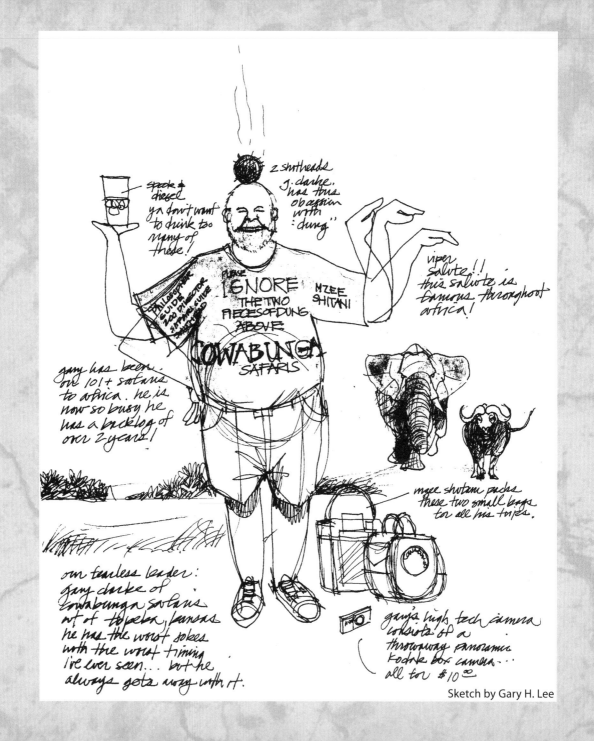

Sketch by Gary H. Lee

Safari P's & Q's (Puns & Quips)—By Request!

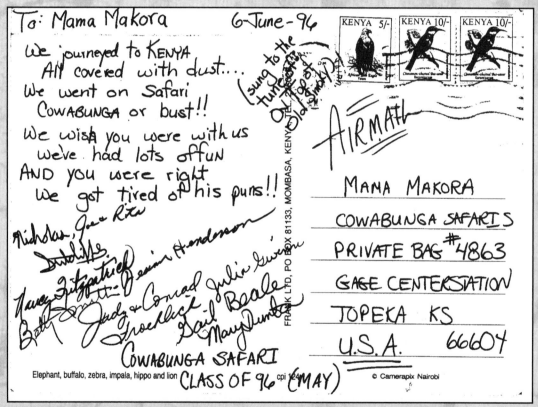

To: Mama Makora   6-June-96

We journeyed to KENYA
   All covered with dust....
We went on Safari
   COWABUNGA or bust!!
We wish you were with us
   We've had lots offun
AND you were right
   We got tired of his puns!!

(sung to the tune of Oh, Tannenbaum ("Old Family))

Nicholas, Joe + Rita
Dutchliff
Nancy Fitzpatrick
Betty Smith, Jerrie Henderson
Judy + Conrad
Froehlich, Julia Gwenn
Gail Beale
MaryDunton

COWABUNGA SAFARI
CLASS OF 96 (MAY)

Elephant, buffalo, zebra, impala, hippo and lion   cpi /24

FRANK LTD, PO BOX 81133, MOMBASA, KENYA TEL 26051

© Camerapix Nairobi

AIRMAIL

MAMA MAKORA
COWABUNGA SAFARIS
PRIVATE BAG #4863
GAGE CENTERSTATION
TOPEKA   KS
U.S.A.   66604

KENYA 5/-   KENYA 10/-   KENYA 10/-

Postcard from a safari group in Africa to Mama Makora ("Mama Trouble"—the KiSwahili name for Nancy Cherry), the Administrative Officer at Cowabunga Safaris Main Camp.

Even after reading the following disclaimer, folks still signed on: *The management and staff of Cowa & Bunga Safaris are not responsible for bad puns on the part of the Safari Leader.*

But after an entire safari of my bad puns and corny jokes, most people were greatly relieved to be done with them . . . and me!

A few, however, had said, "You should do a Safari Joke Book." Knowing full well that such an endeavor could not stand alone, I have included some in this chapter, as well as other humorous incidents.

Many are in the "you had to be there" category, and I've tried to set the stage for those who have not yet been on safari. Some of the jokes are not so funny, and some of the puns are not so punny. Cowabunga Safaris alumni may want to proceed to the next chapter. For those of you who are brave enough to suffer through, get ready to groan. Here we go. . .

囙 囙 囙

On safari, the roughest roads are toll roads.

Really?

Yes, they take their toll on you.

꩜ ꩜ ꩜

The sausage tree is the worst tree of Africa.

Why so?

Think about it . . . sausage . . . wurst . . . .

꩜ ꩜ ꩜

A dik-dik is a very small, delicate antelope native to thick African bush. Its formal name is Richard-Richard.

꩜ ꩜ ꩜

Two old vultures in Tanzania had just finished picking a carcass to the bone. "We need a vacation," said one. "Let's go to Cape Town," said the other. Knowing it was a bit much to fly half the length of the African continent at their age, they booked a flight on a commercial airline. When they checked-in at the counter the agent asked, "Do you have any baggage to check?"

"No," they replied. "Just carrion."

೫ ೫ ೫

Out on the game drive did you see a bat-eared fox? No, but back at camp I saw a fat-eared box.

೫ ೫ ೫

There was a toad sitting on my toilet!
Really? Did that make it a toad stool?

೫ ೫ ೫

One of the more colorful and noisy birds in Africa is the shrike. Americans on safari are not permitted to see more than two shrikes . . . because . . . three shrikes and you're out!

೫ ೫ ೫

While searching for owls, we peered into a hollow tree cavity, but the owl had flown away. That was the first no owl.

೫ ೫ ೫

At camp in the evenings, we often heard a marvelous chorus of frogs. I always cautioned the group to listen closely, as the frogs would all be dead in the morning.

"Why?" they would ask.

Because they are croaking tonight.

𐋡 𐋡 𐋡

Around the campfire one night, we heard two frightening sounds in the darkness: to our left the yelp of a jackal, and to our right the howl of a hyena.

"What was that?" the group asked anxiously.

"Dr. Jackal and Mr. Hyena," I replied.

𐋡 𐋡 𐋡

Elephants love to strip bark off huge baobab trees with their tusks. Once I saw an elephant strip bark off an acacia tree. He *thought* it was a baobab, but he was . . . barking up the wrong tree.

𐋡 𐋡 𐋡

At a remote camp outside of National Parks protection, I requested poached eggs for breakfast. After a lengthy delay, the

head waiter advised: "We are experiencing a poaching problem in the kitchen."

≈ ≈ ≈

A popular small mammal in Africa is the meerkat, also called the suricate.

So . . . would a substitute mother meerkat be a surrogate suricate?

≈ ≈ ≈

I've always admired dung beetles: they put up with a lot of shit, but they've sure got balls.

≈ ≈ ≈

What did the grape say when the elephant sat on it?

It let out a little wine.

≈ ≈ ≈

In hot, dry Namibia there is a huge extinct lake bed called the Etosha Pan. Midday, our group stood at the edge of the Pan in the glaring sun. I told them to have their cameras at the ready, and then I ran onto the pan, removed my hat to expose my shiny bald head, and became . . . a flash in the pan.

卍 卍 卍

On safari, days would start very early, well before the sun rose. As one group stood around the campfire drinking tea or coffee in the pre-dawn darkness, someone exclaimed, "Ah! What a beautiful morning!" To which a dour voice responded, "Looks like last night to me."

卍 卍 卍

Our camp on the equator had a dramatic view of Mount Kenya. On our arrival it was shrouded in clouds, but I indicated its location and told folks to watch for it. Several days later it cleared, and as the group was taking photographs one person asked, "Has it been there all this time?"

卍 卍 卍

Some of my most colorful and enthusiastic safarists were farmers, and many were those salt-of-the-earth people I call root-stock Kansans. I loved their characteristic lingo. They referred to the famed Olduvai Gorge as "oldie gulch," while the immense, green African savannah was "mighty fine pasture," and the various species of wild animals were "a different breed of livestock."

꒰ ꒱ ꒰ ꒱

In order to pinpoint the location of an animal on first sighting, I briefed each group on how to use the "clock system." Straight ahead is 12 o'clock, direct right is 3 o'clock, 6 o'clock is the opposite of 12—to the rear, and 9 o'clock direct left, with corresponding points in between. Of course, I was always looking for "the gnus . . . at 10 o'clock."

We were on a game drive in an area where rhinos had not been sighted for years. One of my experienced safarists—a tall Texan with superb spotting skills—was standing in the back of the vehicle with his binoculars, while I scouted in front. Suddenly he yelled: "RHINO! RHINO! RHINO!" Had it been anyone else, I would have dismissed it as too good to be true, but you never know. I twisted around and glassed the area. Nothing. "RHINO! RHINO! RHINO!" he repeated. Frantic, I said, "What time?"

"Right now!" he said. "Right now!"

I asked which *direction*, and he pointed toward 5 o'clock. Yes, there was a large animal covered in mud moving through the bush!

A rhino? No.

Warthog.

꯱ ꯱ ꯱

One lady claimed to be a connoisseur of wine, and enjoyed a glass or three each evening with dinner, though I never saw her actually read the label. One evening, toward the end of the meal, the waiter whispered in my ear: "Cowabunga! There is a mother rhino and baby at the waterhole!" I signaled to the group to be very quiet and we slowly moved outside. In the dim light, one could readily see the silhouettes of these rare animals. What a thrill!

Squinting her eyes the wine lady asked me, "What are we looking at?" I moved her a few steps so that she had a better—

and direct—view. "It is two rhinos, right there," I said. She looked . . . and looked . . . and finally said, "If I wasn't such a wino I could probably see the rhino."

❀ ❀ ❀

Please note those were *her* words, not mine. I'm in no position to pass judgement about being overserved on sundowners, as you shall see. One of my favorites is a South African tangerine liqueur called Van Der Hum. It is exquisite, and after just a few, it is Van Der *Hummmm . . . .*

The group on one of my Zimbabwe safaris consisted of Topeka Zoo Docents, those wonderful volunteers that extend the educational dimensions of the zoo on a personal level. I had worked with these ladies over the years, and they were quite familiar with my sense of humor. Good vibes flowed between us and in the course of our journey we had a lot of laughs, especially about five-legged elephants. Before I excused myself to retire each night, they <u>expected</u> me to use that old, tired line: "You want to come and see my etchings?" They would dismiss me with a mocking guffaw and send me on my way.

One night after dinner at our lodge on Lake Kariba, the ladies treated me to a Van Der Hum. *Well, that is very kind of you.* Then a second one. *Thank you so very much.* And then another. *Thanks again.* And another. *Uhh . . . Ok.* And . . . *Ladies! Please excuse me. I must turn in. Good night.*

I tottered back to my room, dropped my clothes in a pile, and stepped into the shower. Just as I reached to turn the water on, there was a KNOCK! KNOCK! KNOCK! on my door. What the . . .?

I wrapped a towel around my waist, opened the door, and there in the hall were *all* of the ladies! Then—together in a sing-song voice—they proclaimed: "GAAR-Rie—we-CAME-to-SEE-your-EETCH-ings."

HA! The entire evening had been a set-up and I was royally duped! When the shock wore off, I had to confess to the ladies that, "I couldn't show you my etchings now if my life depended on it."

And again, together in a sing-song voice, they declared: "GAAR-Rie—THIS-is-YOUR-*only*-CHA-ance."

Laughter filled the air as the ladies moseyed down the hall.

ॸ ॸ ॸ

With reference to ladies who have put me in my place, the champion may very well be Julia. At 88, she was the oldest person I had ever taken on safari. I can't imagine that she weighed more than 88 pounds! She was like a little bag of bones, and I do not mean that in a patronizing way. I worried about her! The roads on safari can be *so* rough, the days are long, meals frequently are irregular, qualified medical help may not be available—all of these thoughts and more, were flooding my mind. At our pre-safari briefing and during the flights to Africa, she never had one question or comment about our trip.

On our first morning in Africa, before we hit the safari trail, I explained the social structure of an elephant group, and how the wisest and most mature female was the matriarch. To make sure that Julia felt a part of the safari, I asked her to step forward and said I wanted to designate her as the matriarch of our safari group. She looked up at me and, shaking her finger said, "And don't you forget it, either!"

But that did not deflate my ego nearly as much as an excited pair of young ladies on their first safari. They were a joy—*so* interested in everything, *so* rhapsodic about simply being in Africa. We were in lion country much of the time, and at night we frequently heard the lions' magnificent roars. This was always a topic of conversation over breakfast. Everyone was thrilled—except these two ladies. For whatever reason, they had not heard a single lion roar! I felt bad, and did not want them to miss out on this special dimension of the safari experience.

On the last night of our safari, after everyone had turned in, I quietly stole to the back of their tent and imitated my best lion roar. I could just visualize the ladies bolting upright in their respective beds, maybe reaching out in the dark to touch hands for reassurance. I half expected a muffled scream.

Silence.

Again I roared, louder this time (and I must admit, I'm pretty good at it).

Still silence.

I was perplexed! If they truly *were* sleeping that soundly, no wonder they had not heard the real lions.

I slipped over to staff quarters and borrowed a metal bucket. I knew from experience that it would enhance the resonant sound, more like that of a lion.

Back behind the ladies' tent, I stood in the dark, with the bucket on my head, and let loose with the-mightiest-lion-roar-of-my-life!

From inside the tent, two voices in unison said, "Good night, Gary."

卐 卐 卐

Ladies Behind the Baobab Tree, Gentlemen to the Termite Mound

Where would ladies on safari go to the toilet . . . if there was no toilet?

Through my many years as a safari leader, I had a medley of responsibilities—the safety and well-being of the group, illness or injury, airline connections, lost luggage, overbooked accommodations, leaky tents and boats, broken-down vehicles, impassable roads, bridges or lack thereof, elephants who have right-of-way, and other unexpected eventualities.

As a *male* safari leader I had an additional and overriding concern: how ladies would relieve themselves on the trail or in the bush with no facilities. While this is a shared human biological function, I felt it was a bit more problematic for women. Hence, it was important for me to be sensitive to possible embarrassment, to respect privacy, yet insure protective vigilance—not only against dangerous large animals, but insects, thorns, irritating vegetation, warthog burrows and even hyena dens.

Many modern safari lodges and permanent tented camps have Westernized flush toilets. In the course of a safari, however, one can expect to experience *traditional* safari toilets. Top-of-the-line would be the "long drop," a toilet seat placed over a hole six feet deep, usually with permanent walls and roof, frequently with a half-door in front, so you can enjoy the view but still have privacy. On mobile camping safaris, it is customary to use a "short drop," which is a hole three feet deep with a toilet seat over it, often in a small tent, used on a temporary basis. To "flush," you drop a scoop or two of nearby loose soil from the freshly dug hole back *into* the hole.

One of my favorite short drops was at our camp along the Zambezi River. Instead of sitting on the traditional oval , the lower jawbone of an elephant turned upside down over the hole served the same purpose! And on desert safaris, we used the standard toilet seat simply supported by four metal poles and stuck in the sand, a true "open-air" design *in the open*—no tent, no walls, nothing. Proper safari etiquette ensured that your privacy was respected while you were marking your territory.

During our safari travels, we would encounter Eastern toilets: a room, sometimes tiled, with a hole in the floor—no toilet bowl, not even a toilet seat—imagine the surprise of the lady who brought toilet seat covers. To use it entailed positioning oneself over the hole with a foot on either side, then standing or squatting with nothing to sit on and nothing to hold on to. After her first experience with this, a grandmotherly lady came up to me and said, "You know, Gary, to use that toilet you have to be an expert marksman."

In all of our Cowabunga Safaris verbal and written briefings, we emphasized that each safarist should have an emergency supply of toilet paper with them at all times. Everyone took heed and usually carried it discreetly, folded in their safari vest or camera bag. One lady, however, always had a large roll of T.P. *in hand,* even on the table at meals. She took the advice to heart

and wanted to make sure I knew it.

Incidentally, we ensured that Cowabunga Safarists carried plastic bags to bring used T.P. back to camp for proper disposal. Obviously we didn't want to leave it as litter or burn it, risking a bush fire, or even bury it, as animals might dig it up and ingest it. Plastic bags were also good for holding used napkins or tissues, wrappers, A-B-C gum, moist towelets, or any other litter generated in the usual course of our daily activities or journeys.

Before every safari activity and prior to departure for the next camp or lodge, I would remind everyone to "skip to the loo" even if they didn't have the momentary urge.

I've always liked the quote attributed to Winston Churchill, who when asked how he accounted for his greatness, reportedly said, "Never pass up an opportunity to urinate."

In contrast to a safari toilet, a bush loo means doing your business in the open wilderness. Whenever possible during safari activities, we would make periodic "pit stops" for such a purpose. After a precautionary scouting of the territory, I would designate which gender was to go where, hence, the title of this chapter.

If someone had to go unexpectedly between pit stops, we asked that they give us five or ten minutes notice if possible, so that we would have time to find a safe place with privacy. I had a lot of fun with this in KiSwahili-speaking countries, where the word for toilet is *choo*. The word is not pronounced like "chew," but with a long "o" to rhyme with "show." I would tell the group that if they needed a toilet, they should simply sing out in a rising voice, "i-i-i-t-t-s **CHOO**-Time!"

Some of the ladies came up with helpful hints for their sister safarists. Since it was usually necessary to squat without the convenience of something to sit on, this put a strain on the thighs. A pre-safari exercise (dubbed the *Tanzania squat)* helped alleviate the discomfort. Also, when squatting on an incline, be sure to have your toes pointed *uphill* to keep your shoes dry.

I recall a game drive with five ladies and two men—me and the husband of one of the ladies. The ladies alerted me in unison of their need for a bush loo. We were in a desolate, arid region with no safe cover close to the sandy road. Since we had not seen another vehicle for hours, I suggested that we stop in the middle of the open road, and have ladies go in back of the vehicle and gentlemen in front. All agreed.

It was very quiet—no wind, no bird song—and the husband and I could not help but hear the ladies laughing and chatting. Then all fell silent, except for two distinctive sounds: the tinkle of urine in the sand, coupled with the whining about aching legs.

The husband looked at me and said: "Gives a new meaning to the term 'pissing and moaning,' doesn't it?"

On hot air balloon flights over the African savannah, it is customary after landing to have a champagne breakfast, cooked on the spot. This is always a festive occasion, with champagne flowing.

I remember one lady who requested a bush loo; we checked the area and directed her to a large boulder nearby. As is frequently the case, she asked her tent-mate to go with her. The tent-mate said she would, just as soon as she got the label off a champagne bottle.

"Oh, that's OK," the lady replied. "I've got some toilet

paper."

Ladies seem to be featured in a disproportionate number of my "loo stories," so let's tell a few on the guys. Usually the group was on land when nature called, but on one occasion we were in a boat. Once the bow was secure on shore I announced, "Ladies to the right, gentlemen to the left." One by one safarists carefully stepped out of the boat and turned in the appropriate direction. The last man out to step ashore was hard of hearing, and he proceeded to follow the ladies. I tapped him on the shoulder, mouthed the words 'BUSH LOO' and pointed to the opposite direction.

"Oh," he said with surprise. "I thought we were going on a nature walk."

Here's another one. My tent-mate and I were the only males in a vehicle filled with ladies. It was early morning and we had been out a short while when, with embarrassment, he announced, "Sorry, folks, but I've gotta go—bad! Number two. I can't hold it."

Everyone understood the urgency of the situation, and we

stopped right where we were. And where we were was . . . in the middle of a migration, completely surrounded by wildebeest—thousands of wildebeest—which are also called the gnu (pronounced "nuu") because of their characteristic sound.

My friend made a hasty exit with compressed buttocks out of the vehicle and into the mass herd. What choice did he have? Fortunately, gnus are not aggressive. Staying as close as possible to the vehicle, he moved directly behind it with gnus all around, and the air filled with the continuous hum of, "nuu, nuu, nuu, nuu, nuu, nuu, nuu . . ." ad infinitum.

The ladies and I sat patiently in the vehicle, eyes front, but unable to carry on much of a conversation because of the incessant, "nuu, nuu, nuu, nuu, nuu, nuu, nuu," from the gnus.

Presently my tent-mate returned with a silly grin on his face. "Why," he exclaimed, "this is just like at home—I can do my morning constitution and listen to the gnus at the same time."

One of my most rugged safaris was overland through the sand tracks of Botswana in old Ford trucks, with minimal gear, staying at public campsites. At our campsite in Chobe

National Park, the rules dictated that we use the 'ablution block' for toilets. This consisted of an ill-kept, rectangular concrete structure with no running water, no roof, no windows, and no interior light. Inside were several stalls, each containing a long drop seat.

After we had set camp, I gathered the group for an ablution block briefing. I wanted them to become totally familiar with the path from our tents to the block in the light of day, as I knew it would not look the same at night with just a torch. The path had twists and turns, rocks and tree roots, branches and thorns. The building had no door, which meant animals might be inside. Hyenas were already hanging around our campsite.

I did not like this set-up at all, but felt we should follow the rules, and the group bravely agreed. I offered to accompany anyone, at any time, during the night. Most folks said they planned to drink less after dinner to try to make it through the night. One dear lady, however, *did* have to go several times each night. She would carefully follow me down the path, I'd check inside for animals, then wait outside, and we would return to

the dark campsite. All the while, we'd hear hippos snorting, lions roaring, and hyenas howling. She must have been a bit frightened, but never complained . . . until our last morning in camp.

During the night, an elephant had entered camp and stopped next to her tent. When I came out of my tent at sunrise, she was already standing in front of hers, hands on her hips, intently looking at a huge pile of fresh elephant dung encircled by a giant patch of wet sand.

She looked at me and asked, emphatically: "Gary . . . if an **elephant** can go just **outside** my tent . . . WHY – CAN'T – I?

What could I do but shrug my shoulders?

Lasting memories from a safari might include a rare animal sighting or unusual behaviors, a breathtaking landscape or a dramatic sunset, the night sounds of Africa or sleeping in a tent for the first time. But I dare say that probably all safarists have a toilet or "bush loo" experience they will never forget. This was particularly true for one of my Botswana groups in 1982.

They were an adventurous group, camping along the Chobe

River, and dubbed themselves the "Chobe Chums." At our reunion they presented the "Chobe Chums Awards." Some of the awards pertained to loud snoring, lost cameras, or safari attire, but the best awards were . . . well, read on.

SPECIAL CHOBE CHUMS RECOGNITION AWARDED FOR:

Most Frequent Use of Men's Toilets:

Sally Clark

Standing-in-Elephant-Dung-While-You-Eat-Lunch Award:

JoAnn Myers

Most Pit Stops:

Brenda Strickler

The How-to-Urinate-on-Your-Shoe Award:

Joy Lamb

And we all had a good laugh, again.

All the animals of Africa, of course, go when they need to, wherever they are at the time. The resulting scent marks are checked by members of their own species, as well as various other animals. It's like reading the bush guest register about who has been in the neighborhood. I've often wondered what a lion or an elephant or a zebra might think if they were to sniff the damp residue of my dawn urine at a bush loo by a termite mound. One of my safarists suggested it would be something like this: "Ah, Cowabunga was here, and last night for sundowners he had one gin-and-tonic and a local beer."

The *Compact Oxford English Dictionary* defines *throne* as, "a ceremonial chair for a sovereign." I've heard the word used with reference to the toilet as well, and the best example of this was on an island in the Okavango Delta, an aquatic wonderland in Botswana. An enormous, abandoned termite mound of hard-packed clay had been converted into an elevated "loo/throne." While sitting on it under the African sky, surrounded by the beauty of nature itself and not in the confines of tile, porcelain, or artificial light, one truly does feel like royalty . . . or at least I did.

Be it behind the baobab tree or from the termite mound, the view from the loo is best when in the bush of Africa.

A Proper Proportion of Propp's Popular Productions

The "Propp" in the title is Burton M. Propp, CPA, of Oakland, California. The "production" is a set of his superlative safari cartoons.

I first met Burt and his wife Joan on a hot air balloon flight over the Maasai Mara in Kenya in September 1994. My group filled ten of the twelve spaces in the balloon basket, and this delightful couple joined us. It was their first safari, so to be hospitable and to make them feel welcome, I designated them as "Honorary Cowabunga."

Once aloft, I regaled the group with my usual "attempts" at humor. Since I had already been on more than fifty safaris, as well as several dozen balloon flights, I had a lot of material. Yet each time I cracked a joke or made a pun, Burt responded with

a better one! My witticisms elicited only groans from the group, while Burt was rewarded with hearty laughter.

The group was elated that Burt was flying high (no pun intended) and they egged him on to greater heights (no pun intended). Even the pilot was laughing.

I was gobsmacked!

Seldom—maybe never—had I been so trounced in a match of verbal wits, and by a charming, likeable, unassuming "unknown comedian."

Much to my chagrin, he was funnier than me—*much* funnier!

Upon landing, I acknowledged Burt as the worthy champion and toasted him at our champagne breakfast in the bush.

I invited Burt and Joan to join us on a future Cowabunga Safari. They were part of our group to Zimbabwe and South Africa in August 1997. We had a marvelous time, and Burt was in top form. Within a month after our return, I received a packet from Burt. It contained a series of original cartoons he had produced on his computer that reflected our Zimbabwe safari experience. I keep them at Main Camp for all to enjoy, and they are quite popular at safari reunions. Wanting to share Burt's original safari humor, I sought permission to feature some of his cartoons in this book.

Burt graciously agreed, in a letter dated February 19, 2010 as follows. Keep in mind that the originals are in color. You are welcome to pop into Main Camp at any time to see them.

So, with admiration and appreciation, I proudly present A Proper Proportion of Propp's Popular Productions.

February 19, 2010

C OWABUNGA!

WOW! A VOICE FROM THE PAST, THE INFAMOUS ZOOMEISTER HAS RISEN! IT IS GOOD TO HEAR FROM YOU AND THANKS FOR THE INQUIRY ABOUT JOAN AND MYSELF. WE ARE BOTH WELL AND I JUST CELEBRATEDMY83RD BIRTHDAY AND AM STILL WOR,KING (WELL AT LEAST 4 MONTHS A YEAR) SINCE MY CLIENTS WON'T LET ME QUIT. WE DO A LOT OF TRAVELING THE REST OF THE YEAR

BUT NOTHING AS EXCITING AS OUR AFRICAN ADVENTURES WITH YOU. NOW IT IS MOSTLY CRUISES BUT WE ARE RUNNING OUT OF PLACES TO CRUISE. I WOULD LIKE TO GO BACK TO AFRICA BUT JOAN FEELS AT HER AGE SHE CAN NO LONGER CLIMB IN AND OUT OF THE ANIMAL VIEWING TRUCKS AND THE RIGORS OF PACKING, UNPACKING AND GOING TO A NEW PLACE EVERY COUPLE OF DAYS. MY MOST MEMORABLE AND EXCITING VENTURE WAS OUR LAST TRIP WITH YOU TO ZIMBABWE. I WILL NEVER FORGET STANDING OUTSIDE MY TENT AND BEING ABLE TO REACH OUT AND TOUCH A WILD ELEPHANT PASSING BY ON HIS WAY TO THE RIVER.

I HOPE YOU ARE DOING WELL AND I GUESS YOU CONTINUE TO DO YOUR SAFARI'S? MY JOKESTER BUDDY GEORGE SHEARER PASSED AWAY 4 OR 5 YEARS AGO; HOWEVER I AM LUCKILY STILL HERE AND STILL HAVE MY REPERTOIRE OF JOKES, SHOULD YOU NEED SOME. I FEEL VERY FORTUNATE TO HAVE MET YOU ON MY FIRST TRIP TO AFRICA AND CONSIDER CANOEING THE ZAMBEZI WITH YOU AS THE HIGHLIGHLIGHT OF ALL MY VARIOUS TRAVELS.

BEST REGARDS, BURT & JOAN

AFRICAN MAMMAL IDENTIFICATION
CAPE BUFFALO
(Syncerus caffer)

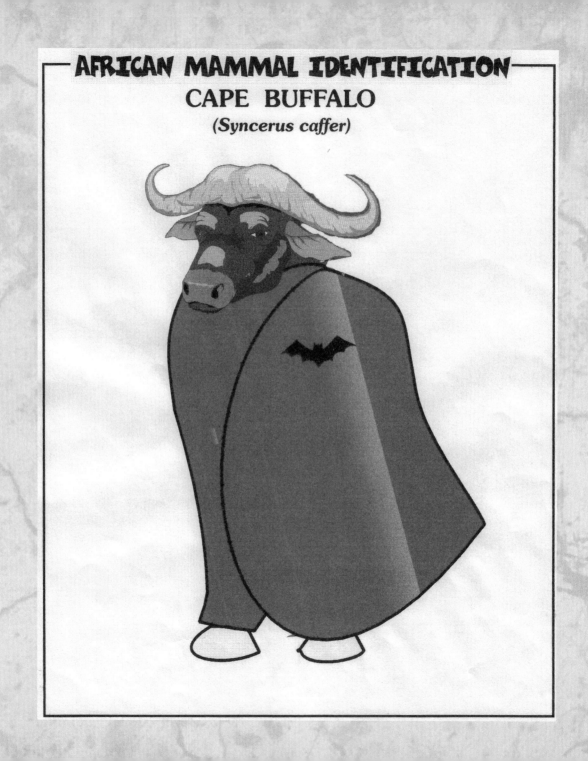

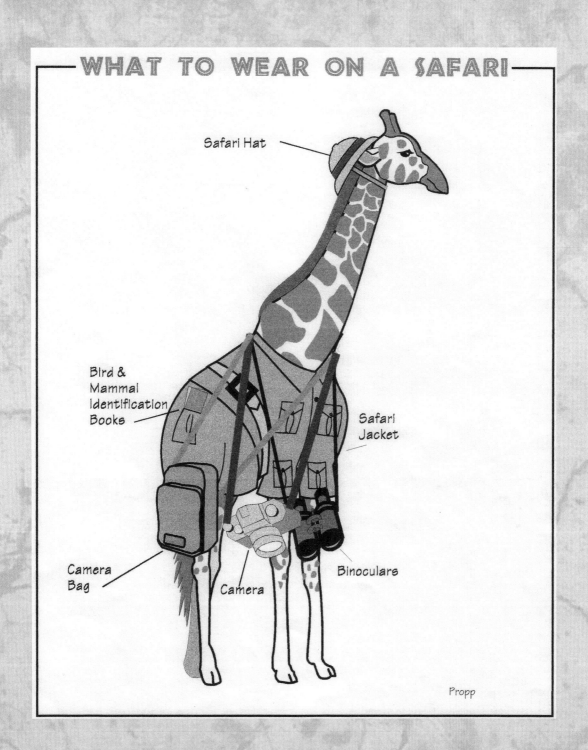

AFRICAN MAMMAL IDENTIFICATION

SPRINGBUCK

(Antidorcas marsupialis springis)

propp

Oh no! Not leftovers again!

AFRICAN MAMMAL IDENTIFICATION

Wild Dog
(Lycaon pictus)
(With draftus brewus)

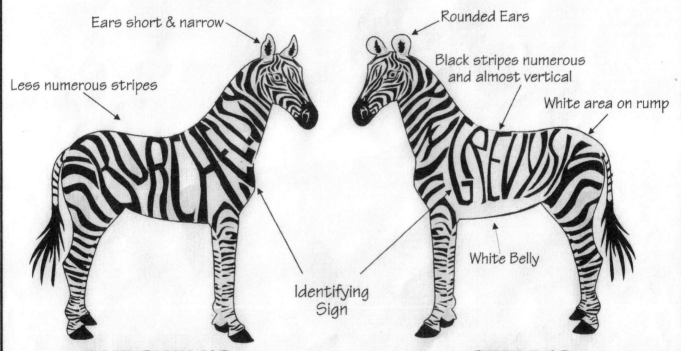

HELPFUL HINTS FOR YOUR SAFARI

(FOR THOSE WHO CAN'T TELL A GREVY'S ZEBRA FROM A BURCHELL'S)

Ears short & narrow

Less numerous stripes

Rounded Ears

Black stripes numerous and almost vertical

White area on rump

Identifying Sign

White Belly

BURCHELL'S

GREVY'S

Propp

Hello, Home Office, I thing we have a glitch in our computerized painting program, the animals are coming out mismatched

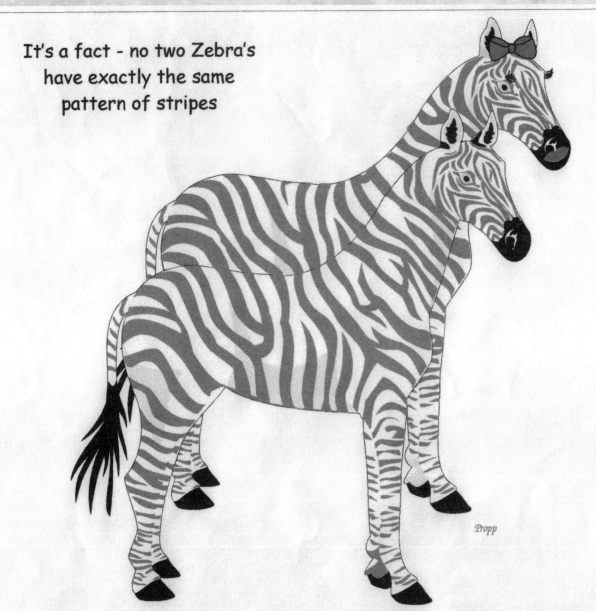

AFRICAN ROAD SIGNS

CROSSSING

CAUTION
WATCH YOUR STEP

propp

In Search of the Perfect Bolus

It was all Garth's idea.

He approached my group behind my back while on our safari to Tanzania in January 1986. At first they were shocked, then stunned—but they also realized it would be an effective way to "pay back" all the pranks and bad puns I had inflicted on them.

A bizarre notion, yes—but once the group got used to it, they decided that *if* they pulled it off, the end result would be well worth it.

Long-time friend Garth Thompson, my canoeing guide on the Zambezi River, knew elephants—from trunk to tail . . . and everything in between.

Garth's brainchild was to make a birthday cake for me from an elephant bolus. Fortunately, fresh elephant dung looks and smells like wet hay, as it is primarily undigested grasses and leaves. The conspiracy took a great deal of secrecy between Garth, the group, and the drivers, with some rather unusual game drives as a result.

Garth explained that the bolus had to be shaped like a birthday cake and had to be just the right consistency. Their assignment for the next several days was to search Tarangire National Park for such a treasure, without me knowing. That meant that anyone riding with me could not be obvious in their search for a perfect specimen. Those in other vehicles could seek with zeal and vigor—looking for elephants, and consequently boluses, to find "just the one" that would please Garth.

After each game drive, as we all returned to camp, a "decoy" offered me a beer and asked lots of questions to keep me occupied. The rest of the group and the drivers unloaded the vehicles as soon as possible (sometimes the boluses were <u>too</u> fresh) and laid them out on the ground. Garth then inspected each one and if he wasn't happy they'd have to do the same drill on the next game drive. The rejected boluses were surreptitiously dumped just far enough away from the camp to not arouse suspicion. Actually, elephants did come very close to camp on a regular basis.

Garth finally found the bolus of his dreams and carefully hid it so I wouldn't see it. Early on, Garth had met Annette, who managed our lovely tented camp, officially known as Tarangire Safari Lodge. She was very understanding when Garth explained the joke to her, and having known me for several years, she agreed that I was the perfect foil for this bizarre birthday cake. There were just two problems.

First, the "cake" needed to be iced to look bona fide. Annette sadly informed Garth that her supplies from Arusha had not yet arrived and sugar was a critical camp commodity at the moment. But not to fear, for Annette saved the day. During the time we were at her camp, the staff had been whitewashing the rocks that bordered the entrance drive. So when the group departed on the next game drive, she had a staff member whitewash the bolus!

Second, and somewhat more difficult, was that my birthday did not occur at Tarangire. It was actually several days later when the group was at Ndutu Tented Camp, in the southern

Serengeti. The "cake" needed to be kept cool while in transit to preserve the shape. Cool boxes (similar to our picnic coolers) were a precious item in Tanzania at that time. But after all they'd been through, and all she had done, Annette couldn't let the group down now. She loaned one of her special cool boxes to transport the cake to the Serengeti. Our head driver made sure it was hidden *under* the luggage in his vehicle so I wouldn't see it. Mind you, all of this was done in deepest secrecy over the next three days, en route.

Often I sense when I am about to become the target of a practical joke. But not this time: in addition to habitually insuring that my group was hale and hearty, I was busy with animal sightings and behaviors, making journal entries and wildlife checklists, taking photos, meeting people, and improving my KiSwahili. I did not have a clue.

We arrived at the Serengeti in late afternoon, a bit-travel weary but in good spirits. Garth immediately advised the Ndutu camp staff—I'm sure they were delighted to be in on the prank. After sprucing up, we gathered around the campfire to relax with sundowners, and to recount our adventures thus far. Dinner was by lamplight with animated conversation. I still had no clue.

Dessert was served and then—SURPRISE! My group started singing "Happy Birthday" and the camp staff danced around the table in a conga line. The leader was carrying a rounded birthday cake (no kidding) covered in white icing (lovely) with a single lit candle in the center. He placed it in front of me, I made a wish (that I would return to Africa), and I blew out the candle. The group applauded enthusiastically.

I was presented with a sharp knife to cut the cake. Quite candidly, it did not seem large enough to serve the entire group, but I proceeded with the ceremony. Wow, what a tough cake . . . very fiberous . . . and funny looking inside . . . why is everyone laughing and taking pictures?

Of course, the group could not wait to tell me the whole story, and I was honored they went to so much trouble. They blamed it all on Garth, and rightly so. However, it did start a tradition, as an elephant bolus cake has been served to numerous Cowabunga Safarists since then.

On 1 March 2013, I received the following note from Garth:

Hi, Cowabunga. You cannot believe how shocked and delighted I was to see the same couple running Tarangire Safari Lodge after all these years. Remember, we stayed here in 1986 and they helped me make the only dung cake ever made with whitewash for icing!!! Remember, we delivered it to you on your birthday in the Serengeti!!! Best, Garth.

Yes, Garth, I remember, and I'm glad you are *still* excited.

Celebrating a birthday on safari was always special to me, and some of the gifts were particularly noteworthy: customized tee shirts, fashionable African dress shirts, art prints, books on Africa signed by the entire group (and sometimes the author), wood carvings, bronze sculptures, hand-tooled leather place mats, camp stools and travel bags, as well as other treasured keepsakes. And then there were those gifts that were "just for the fun of it."

I recall one instance at a remote bush camp when, unknown to me, the group decided that all gifts had to be handcrafted solely from available natural materials. It was a creative group, and many of them produced items from feathers, woven grasses, flowers, dried leaves and seed pods. One lady carefully collected the dung pellets of the greater kudu antelope and skillfully strung them together for a striking kudu doo-doo necklace.

One fellow got the last laugh on me, to the delight of the group. He took some long, flexible tree branches and fashioned them into the exact size and shape of a tennis racket frame. Upon presentation, I jumped up and pantomimed my best tennis moves (HA!). Of course, I had to chide the craftsman a bit, so I stopped and exclaimed: "Wait a minute! This is a tennis racket with no net." With a slight smirk he confidently replied: "It is for a man who has no balls."

丩 丩 丩

Close To My Heart

Larger Than Life

The elephant walks silently through the African forest.

An adult male, he is clearly lord of his domain. Standing over ten feet at the shoulders and approaching five tons, in the lingo of the African bush guides he is a "classic bull."

Emerging from the shadows into a clearing, the sun touches his ivory. His tusks are thick at the base. Both are broken, with fractured, uneven tips, likely long ago in territorial disputes. At one time, they must have been magnificent, but now their broken status gives a rugged dimension.

Pausing with his trunk in the air to test the wind, he shakes his head and changes direction. Well-known and easily recognized in this remote region of the Okavango Delta of northern Botswana, this bull has a name: Abu.

Proceeding at a deliberate pace, Abu heads straightaway to the swamp. He plods into the crystal-clear water thick with reeds and carpeted with water lilies.

Abu does not lift each foot and step into the water. With his great strength he *pushes* through it, creating powerful waves that break the water and sparkle in the sunlight.

The local Tswana people pole their way through the Okavango Delta in a dugout canoe called a mokoro, but here the vegetation is much too thick. Special safari vehicles drive through shallow areas of this swamp, but here the water is much too deep.

Abu is going where only *he* can go.

By now the water is up to the bottom of his ears.

With each step, every muscle flexes. His pelvic girdle sways.

His shoulder blades protrude, moving like giant pistons . . . left, right, left, right.

The thick massive skin on his back rolls, from side to side, across his spine.

I know.

I am riding on top of Abu.

And I feel like I'm riding on top of the world.

റ്റ റ്റ റ്റ

Stranded in Samburu with the Wajukuu!

In November of 1974, I had my largest group ever on safari—24 people. Never again! In June 2001, I had one of my smallest groups—just three. They were my oldest grandchildren, called *wajukuu* in KiSwahili, ages 10, 11 and 11. Never again! Just kidding.

Actually, they kept me as busy as a group of 24, or even 12, but in a different way. They were *ready*, they were *excited* and they knew too much from the start. However, that was my fault.

I usually do not take youngsters on safari until they are at least twelve years old. Of course, when they are <u>your</u> grandkids, you make an exception. Nonetheless, I felt it was important that they be properly prepared for this once-in-a-lifetime experience. So, I devised a course of study called *Safari 101* and a series of *Indabas* (a Zulu term from South Africa meaning "a matter for discussion"). Assignments included African geography, African wildlife, cultures,

KiSwahili, map studies, geography flash cards, and wildlife videos. Additionally, I structured a series of field trips to area museums and zoos so they could hone their mammal and bird-spotting skills, use of binoculars and cameras, and maintain their own Journal and Wildlife Checklist.

The first order of business was to assign a KiSwahili name to each grandchild, based on their favorite or appropriate animal. Kyle, age 11, was *Twiga Mtoto (*giraffe child), Courtney, age 11, was *Duma Mtoto* (cheetah child), and Drew, age 10, was *Nyati Mtoto* (Cape buffalo child).

The weekly Indaba sessions began a year and a half before our scheduled departure date. Sometimes we had them at Main Camp, in my home library/study, or outside under a tree. Our field trips took us to Kansas zoos in Topeka, Manhattan, Salina, and Wichita, as well as Kansas City, Missouri. We visited the Martin and Osa Johnson Safari Museum in Chanute, Kansas, and had numerous bird watching sessions. Just as important, we took a round-trip flight from Kansas City to Chicago so they could learn the nuances and frustrations of air travel.

I'll have to say that the grandkids were eager and enthusiastic, sometimes even correcting <u>my</u> KiSwahili. In fact, their parents often lamented that they wished the children worked this hard on their regular studies! While she was not old enough to actually go on Safari, we included Kyle's younger sister Becca, age 8, in the activities, and she did very well. Her KiSwahili name was *Simba Mtoto* (lion child).

At long last, the big day arrived and the journey began from Kansas to New York, then a 14½-hour flight direct to Johannesburg, South Africa. After an overnight in Joburg, we had a flight up the continent to Kenya, with a spectacular view of Mt. Kilimanjaro above the clouds. On my entry card for Kenya I listed *Babu,* KiSwahili for grandfather, and the immigration official just smiled and waved me through.

There were many highlights from our adventures in Kenya, Zimbabwe (Victoria Falls), and South Africa. Still, one stands out in my mind and that of my *wajukuu* as well.

We were on a game drive in Samburu National Reserve, along the Uaso Nyiro River, watching elephants in the bush and in the river. In our concentration we did not realize that eventually our vehicle was surrounded by **24** elephants. AWESOME! They were so close, yet completely undisturbed by our presence. We had a front row seat, so to speak, to observe elephants in various behaviors—eating, drinking, mud bathing, social interactions—and listen to their sounds—trumpets, rumbles, squeals, even a roar. The grandchildren were delighted, and I was so happy for them.

John, our Driver Guide, felt it best to move down the trail a bit, and that's when we discovered we were stuck in the sand! We dared not get out and push, so John used his two-way radio to call for help.

The only problem was that his frequency was not the same as the nearby Samburu Serena Lodge where we were staying. Hence, he had to radio more than 200 miles to the Nairobi office of our safari operator, to ask them to telephone Serena headquarters in Nairobi to have them radio back to the lodge in Samburu, who then received the message and a rescue vehicle was dispatched. In the time it took for the rescuers to arrive, the elephants moved off and we were able to get out and push ourselves free.

When John radioed Nairobi, that transmission was monitored

by every other vehicle on that frequency throughout Kenya. And when the Nairobi Serena headquarters radioed their Samburu Lodge, that transmission was monitored by every other Serena Lodge throughout Kenya. And, of course, all the other Drivers and Lodge Staffs told everyone else. Unbeknown to us, word spread quickly through this East African nation that straddles the Equator. And everywhere we went on the rest of the safari the Kenyans were smiling and laughing and saying, "Oh, Cowabunga; we heard about you getting stranded in Samburu with your *wajukuu* while surrounded by 24 elephants!"

At first I was a bit embarrassed, but the grandchildren thought it was cool. They were famous in Kenya.

Personally, I liked being surrounded by elephants. And if I'm going to be stranded in Samburu, I want to be with my *wajukuu*.

卐 卐 卐

Authors note: For a sample of a Wildlife Checklist, see the article from *The Just Now News*, 1992, Second Edition (sample on next page).

卐 卐 卐

Asante sana to our underwriters for this issue of Just Now News: Eva Victoria (Queen of Zambezia); Mama Kahawa; Daktari Watoto.

COWABUNGA
SAFARIS
Africa under the rainbow

Printed on recycled paper.

The JUST NOW News - A now-and-then Newsletter for alumni and friends of COWABUNGA SAFARIS published whenever we have enough news and time to put it together.

The **JUST NOW News** 1992 SECOND EDITION

Wildlife Checklist
by Gary K. Clarke

"Well, did you see many animals?"

We see vast quantities and many species of animals on all of our Safaris, more on some than on others. In fact, unless you keep a record, one does not realize the abundance of wildlife you experience.

To give you an idea I've compiled a composite checklist of just the predominant mammals sighted on my three Safaris to Kenya in 1992. This does not consist of every species or subspecies (for example: giraffe actually includes reticulated, Rothschild's and Maasai giraffe). Nor does it begin to cover birds (a considerably longer list).

With mammals, you may see their spoor (dung or track) or hear their sounds before you actually see the animals, or you may see them in special circumstances (at night, or at a waterhole). I've included a few columns to indicate such field situations.

This checklist represents only types of mammals, not their numbers. In some instances we saw only one or two individuals during an entire Safari (e.g., klipspringer), yet during other sightings hundreds or even thousands (wildebeest in migration) were seen. Some sightings are quite special (and lucky) because the animal is rare and endangered (black rhino), secretive and nocturnal (leopard), or restricted to just one habitat in Kenya (sable antelope in Shimba Hills). And, of course, a checklist does not begin to convey the thrill and excitement associated with sighting animals in the wild.

While this list is by no means complete, it gives some idea of what mammals we see on Safari. So, if you want to see animals — lots of animals — join us in Africa.

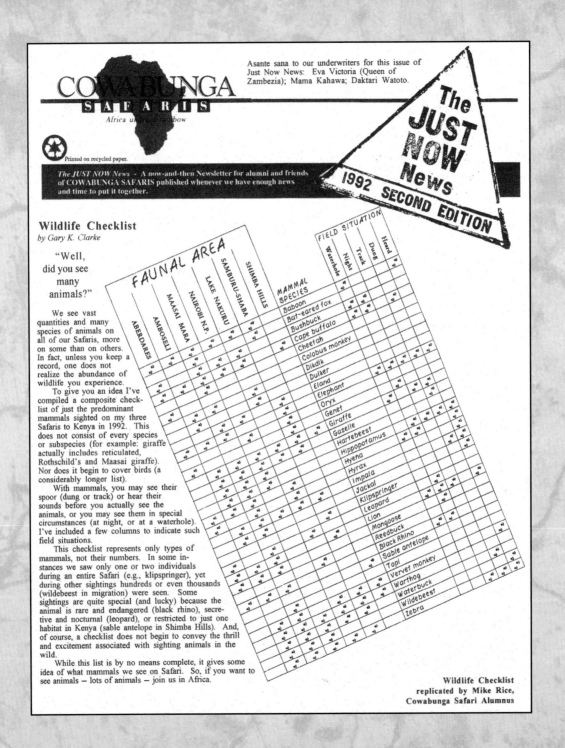

Wildlife Checklist replicated by Mike Rice, Cowabunga Safari Alumnus

Mr. Rhino—Solo

If— I could have just one brother, it would be Steve.

I do . . . and it is.

Although six years younger than me, Steve is not really my "little brother." He is taller and bigger, but that old adage, "He ain't heavy, he's my brother," does *not* apply to Steve—least not the last time I tried to carry him piggy-back for a funny photo.

Beyond our childhood, we have shared a medley of common interests: hot sauce; humor; music; haunting old book stores; offbeat Texas beers; stories about grandkids; chocolate; and philosophizing about life in general.

Our strongest bond is a passion for animals, especially elephants, and zoos—we both worked at the Kansas City Zoo and the Fort Worth Zoo, but not at the same time.

Hence, you can imagine my excitement when the opportunity came to share Africa with Steve on safari to Tanzania in November 1978. I had described Steve to the Tanzanians, and

they were excited that "Cowabunga's brother" was coming to Africa. Because of his size, they dubbed him with the KiSwahili name *Bwana Kifaru,* Mr. Rhinoceros.

Our group members were from New Jersey, Virginia, Kansas, Oklahoma, and Texas. We planned to meet in Chicago and connect with our British Airways (BA) overnight flight to London.

Steve, then Curator of Mammals at the Fort Worth Zoo, departed from Dallas-Fort Worth on Braniff Airways (BN). By the time his flight approached Chicago, strong storms had developed and all arrivals and departures were delayed. My hope was that once the weather cleared and delays unraveled, Steve would still be able to make the BA flight to London with the group.

Whenever I had a far-flung group such as this and we all had to meet at a designated airport for an international departure, my biggest dread was that someone might miss the connection. And now it looked like it might be my brother!

His BN flight was scheduled to arrive at 7:00 p.m. Our BA

flight to London was scheduled to depart at 8:30 p.m.—normally enough connecting time. Then his BN arrival was delayed to 8:00 p.m. But our BA departure was rescheduled to 9:00 p.m.

Updated arrival and departure times were announced every few minutes, and I was wearing myself out running back and forth between the Braniff Airways domestic gate and British Airways international gate, trying to keep up with the changes and insure that Steve made the connection.

Then the BN arrival was delayed further to 8:40 p.m. It looked like Steve would *not* make it. BN and BA both said nothing could be done.

Oops! Hold on! Now Steve's BN flight won't arrive until 9:00 p.m., exactly when BA departs. BA says I must board the flight. But, it is delayed another 13 minutes, so there is still hope. I run back to BN. No word. Oh, no. BN arrival now moved back to 9:20 p.m. It looks hopeless. I run back to BA and leave a message for Steve. HA! I'm the last one to board the BA London flight and the door shuts behind me.

I slumped in my seat exhausted. As we pushed back, I couldn't help but think that my brother was in a bind. Such a mess with all flights out of kilter. The worst part was not being able to talk with Steve, or even leave a reliable message or directions. This had certainly ruined the start of what was to be a glorious safari. I couldn't help but wonder what might happen now . . . .

On arrival at London/Heathrow Airport, I went to the British Airways communications center, where some very courteous staff members assisted me in trying to locate Steve. London time was six hours ahead of Chicago time, so it was just past 5:00 a.m. in Chicago. A total of four telexes were sent to the USA, but by the time we had to board our flight to Kilimanjaro International Airport in Tanzania, we still had no word.

On the long overnight flight from England to East Africa, my mind was in turmoil. I had no idea where my brother was. Still in Chicago? In flight to London? Headed back to Texas?

This was about the worst thing that can happen to a safarist, especially a first-timer. And it was one of the worst things that can happen to a safari leader as well. I felt so helpless and frustrated. I couldn't enjoy the flight, or the fact that I was on

my way to Africa. I didn't even feel like reading. All was dark and quiet in the cabin, with only a faint, eerie blue light glowing from the ceiling. Outside it was clear, with moonlight reflecting off the wing. Every now and then, lights of a city pierced the darkness 33,000 feet below.

I wondered where Steve was, and what he was doing . . . .

We arrived in Tanzania the next morning and proceeded from the airport to Arusha town. Once the group settled in I tried vainly to track Steve's whereabouts. In 1978, communication between Africa, England, and the USA was difficult at best, and my overseas calls produced mostly static, but no new information.

We overnighted in Arusha and the next morning prepared to depart for the bush. Once on the safari trail, I knew I would be out of contact for quite a while. Should I go with the group, or stay in Arusha and await news on Steve? I felt an obligation to be with the group, so I trusted my contacts in Arusha to do all they could. A major concern was the fact there were very few direct flights from Europe to Kilimanjaro.

In the hope that somehow Steve would eventually find his way to Arusha, I left an encouraging note, a map of our routes, and a half-eaten chocolate bar.

From Arusha we proceeded west across the Ardai Plains past Tarangeri National Park, and traversed the floor of the Great Rift Valley to Lake Manyara. Then we ascended to the plateau, through the village of Karatu, and on to the Ngorongoro Crater. After a visit to the Olduvai Gorge and its quaint museum, we continued on to the southern Serengeti Plains, and our lovely tented camp at Ndutu.

Our days were full, the group was happy, but my brother was always on my mind. At each stop along the way, my Tanzanian friends asked, "Where's Bwana Kifaru?" I would relate Steve's predicament. They all were compassionate and reassuring. "Oh, Cowabunga," they would say. "Bwana Kifaru is strong, and he *is* YOUR brother. We will look after him and he will catch up. No problem." And I would leave a half-eaten chocolate bar for him. It was very much like something out of Hansel and Gretel, safari-style.

Nearly a week into the safari, we were around the log fire at our Serengeti camp, the savannah veiled in darkness. Two dim lights appeared from the east. "Cowabunga," the camp staff shouted. "It must be Bwana Kifaru," and so it was. Such a relief!

What a celebration: sundowners and popcorn by the fire, then a joyous dinner with singing and dancing by the staff.

We all were eager to hear from Steve and, tired as he was, he related his tale of misfortune. When his flight finally landed in Chicago, he ran to the international terminal and the BA gate. He saw the London flight sitting on the tarmac, but the agent said there was nothing he could do, so the plane departed, leaving Steve and 50 other passengers stranded. The BA gate agent suggested that Steve take a later flight to Frankfurt and connect on to Addis Ababa, the capitol of Ethiopia, between Sudan and the Red Sea. Apparently the agent knew little of African geography and travel, because he then told Steve he could rent a car in Addis Ababa and drive to Tanzania.

Normally of mild demeanor, Steve exploded.

"Are you CRAZY? NO!"

The agent, no doubt shaking in his boots, issued Steve a hotel voucher and said to come back tomorrow. He did and took an overnight flight to London. At Heathrow the following morning, a BA agent tried to sort things out, but Steve said he spent most of his time staring at the ticket, shaking his head, and saying "What a bloody mess" over and over. The agent could do nothing that day, so Steve overnighted in London and the following day he flew to Brussels, Belgium. That night, on Sabena World Airways, he departed for Africa, very tired of all the hassle and with a severe case of diarrhea.

The aircraft was a Boeing 707 with the front half all cargo and passengers in the rear section. The next morning my brother was at long last in Africa—but not in Tanzania. For whatever reason, Sabena had landed in Bujumbura, the capital of Burundi, south of Rwanda. There was a great deal of civil and military strife in the region. Armed guards escorted passengers to the terminal, where they could use the toilet but could not buy anything.

Remember, while all of this was happening, I'm already on safari with the group, but not really having a good time, because

I'm worrying and wondering where Steve is and how he's doing. Never could I have envisioned the above scenario. Nor could Steve! He thought to himself, "They can put a man on the moon, but they can't get me to Tanzania."

After the unexplained delay in Bujumbura, Sabena continued the flight to Tanzania, but overflew Kilimanjaro International Airport, his destination, continuing on to Dar Es Salaam, then the capital city, on the coast of the Indian Ocean. Steve was advised that a planned connection onto Kilimanjaro was squelched because the Tanzanian Army had confiscated all of the aircraft to use in the border conflict with Iddi Amin in Uganda. So his last leg of air travel was aboard an East African Airways plane to Kilimanjaro. All of the other passengers were locals, including a lady with a chicken! On landing, a military guard inspected his luggage and took all of his hard candy. Steve did not protest, since the guards were armed with AK-47s.

My contacts in Arusha had gotten word of Steve's arrival and dispatched a driver to meet him at the airport. By now night had fallen, and the area was under a military blackout, so he rode for more than an hour in complete darkness—not even with headlights! He could not believe how the driver found and stayed on the isolated road in the dark. Later, I told him that was classic T.A.B.

At the hotel in Arusha, he received a royal welcome by candlelight. "Karibu, Bwana Kifaru—Cowabunga's brother." And he was presented with the first half-eaten chocolate bar.

The next morning he followed this tasty trail over the Ardai Plains, across the floor of the Great Rift Valley, past Lake Manyara and up the escarpment, around the rim of the Ngorongoro Crater, past Olduvai Gorge, and finally into the Serengeti and our camp that night.

Such was the adventure—even before he was truly on safari— of my Bro, Mr. Rhino—solo.

〰 〰 〰

Hippos on the Lawn
Reflections of Rubondo Island Camp

By mid-afternoon, Africa's Lake Victoria extends gray to the horizon and melds with the overcast sky. Strong winds create giant rolling swells that crash white-capped against the rugged rocks by camp. Shy sitatunga antelope peer from beneath the trees. Palm leaves rustle high above our tents.

After nightfall, the sky clears and brilliant stars peek through the forest canopy. Bats flutter in silhouette like imitation butterflies as a genet makes a timid appearance by my tent. The moon, yellow and full, rises through streaked horizontal clouds and casts a rippled reflection on the lake.

At midnight, the wind stills. The only sound is the lapping of water on the beach. Large dark blobs slowly slide across the lawn like small blimps—hippos grazing.

All is quiet until a gentle rain descends, eventually fading to the sounds of slowly dripping vegetation in the darkness.

The humid air carries a potpourri of distinctive odors: tropical flowers, damp vegetation, wet canvas, hippo dung, fish entrails, campfire embers.

With an imperceptible tinge of first light, the piercing cry of the fish eagle echoes across the lake, announcing a new day on the island. Dawn ascends with a chorus of bird calls: African grey parrots, hadada ibises, red-chested cuckoos, laughing doves . . . .

As the sun rises, the lake glows pink and gold. Cormorants and kingfishers dive for food. Giant crocodiles slide into the water. Open-billed storks soar high overhead. A bull elephant pauses along the shore to drink. A stately goliath heron freezes on a low branch, primed to stab a fish with its stiletto beak. Vervet monkeys scamper up palm trees. Spot-necked otters playfully swim past camp. Tiny red-billed firefinches and large black-and-white casqued hornbills venture near the tents.

Golden sunlight filters into the vivid green forest.

This is the paradise called Rubondo Island Camp.

卍 卍 卍

"You Have Been Lost For So Long"

"Where in the hell have you been?"

That is the salutation I might receive in the USA if I show up at some place where I have not been for a while.

Under the same circumstances in Africa, the common greeting is, "Cowabunga, you have been lost for so long! We *know* that you *know* where we are. We know that you love us. And since you have not been here for so long, we know that you must have been lost." Such a lovely phrase, so gracious and endearing, a wonderful expression of humanity through African eyes. You are given the benefit of the doubt, no questions asked.

Another example of cultural civility is the KiSwahili word, *HODI?* This is spoken prior to entering a room, even if the door is open, and means, "May I come in?" I find the African perspective so agreeable.

❧ ❧ ❧

Many Africans live with no electricity and must make productive use of every minute of daylight time. They will leave home in the predawn morning, not to be seen again until after dusk in the evening. Hence, a family may say "Goodbye" at sunrise and "Welcome" at sunset.

❧ ❧ ❧

For freedom of movement, Maasai men traditionally wear a free flowing toga or a long blanket called a *shuka,* simply draped or tied at the shoulder. They wonder why people in Western cultures wear such an impractical and restrictive garment as trousers, covering the body from waist to ankles with a separate part for each leg. They refer to these absurd people as, "Those who contain their farts."

❧ ❧ ❧

Prior to taking a group to a new area in Africa, I made it a point to check it out on my own in order to resolve any logistical problems and to meet the local people along the way. On my first venture into Uganda, I stayed overnight in the town of

Masindi, at the Masindi Hotel, which was one of the original railway hotels—but railway service had been discontinued years ago. Despite its age, the building was in reasonable repair and had a certain colonial charm. The staff was most accommodating and I felt it would be quite adequate for my groups as they were going to and coming from Murchison Falls National Park.

On a subsequent safari, en route to the Falls, I arrived in the evening with my group and the manager rushed out to tell me in an apologetic voice: "Oh, Cowabunga; we have hot water . . . but no electricity." I assured him it was no problem for Cowabunga, and advised the group that they could have an enchanting bath by the flickering light of an kerosene lamp.

Three days later, on our return from the Falls, we arrived in the evening to again overnight at the Masindi Hotel. And again the manager rushed to meet me, but this time he told me with great anxiety: "Oh, Cowabunga; we have electricity . . . but no hot water." T.A.B.

ꄱ ꄱ ꄱ

After my 100th Safari in the year 2000, many of my safari

alumni told me that I should write a book. Hence, I reviewed my journals and wrote a series of short stories about my experiences. Published in 2001, it is titled *I'd Rather Be On Safari,* and has been well received at home and abroad. The African book-signing launch took place in Nairobi at the largest bookstore in East Africa. It was a grand occasion—even Miss Kenya was part of the festivities. The book is quite popular throughout Africa, particularly in the safari camps and lodges. It retails for $26.95, but is marked up to over $45.00 in some shops (which has not helped me one bit).

A year or so after the launch I was in Tanzania on the rugged, dusty trail to the Serengeti when an oncoming vehicle flashed its headlights—a signal to stop. The driver had recognized my vehicle from a distance, and when he pulled alongside, I was delighted to see an old friend—a native Tanzanian safari guide I'd not seen for several years.

"Cowabunga!" he exclaimed. "You have been lost for so long."

We chatted in the middle of the road about family, mutual

friends, recent safaris . . . and my book. "Oh, Cowabunga," he said seriously. "I love your book. It is the best, but it is *soo* expensive!" At that point I wished I had a copy with me that I could give to him.

Then with a smile he continued, "But I know the girl who works in the gift shop, and when she is able, she's making a Xerox copy of each chapter for me." I was honored.

卪 卪 卪

My most unusual Thanksgiving Day occurred in the Northern Frontier District of Kenya in 1976 at a remote camp called Eliye Springs on the shore of Lake Turkana. Formerly known as Lake Rudolph, it is also called the Jade Sea because of the color of the water. It has several inlets but no outlet, and is the largest alkaline lake in the world.

I had briefed the group that Lake Turkana would be a different dimension of Africa, and not to expect elephants, zebras, wildebeest, or giraffes—just camels, sand, desert, and the stately Turkana people. The Lake is famous for its Nile perch, some weighing over 250 pounds, and I advised the group to anticipate fish for every meal: fish cakes at breakfast, fish chowder at lunch, and fish fillets for dinner. Also, I had convinced them it would be truly unique *not* to have a turkey dinner on Thanksgiving, but to dine alfresco, barefoot in the sand, feasting on a fresh-caught Nile perch and looking for the Southern Cross while watching a yellow-winged bat in flight.

But the hospitality of my African friends was far more than I anticipated. The manager knew of our American holiday and the traditional dinner fare. At considerable expense he had chartered a small plane from Nakuru town in central Kenya to fly a turkey to Eliye Springs, a round trip distance of over 500 miles! Much to our surprise, we were served a turkey dinner with all the trimmings. The staff obviously was so pleased that none of us had the heart to tell them how we had been counting on Nile perch for Thanksgiving.

〜 〜 〜

Cynthia Moss and her team of field researchers have studied the elephants of Amboseli National Park in Kenya for over three decades. She has published technical papers and popular books, and her work has been the subject of several PBS-TV documentaries, most recently "Echo: An Elephant to Remember."

At first, Cynthia had difficulty distinguishing one elephant from the next, but soon determined that an elephant's ears provided enough differing characteristics—size, shape, holes, tears along the edges, even the vein pattern—to identify individuals. So with each elephant she met, took a close-up photo of the right ear, a close-up photo of the left ear, as well as a head-on shot of both ears extended forward, and made an I.D. card file to carry in her Land Rover for quick reference. Now she can identify nearly 1,000 individual elephants.

Some years back, I had the privilege of spending a week with Garth Thompson and Cynthia during a lecture tour in Canada. *I learned so much!* In time, she returned to Kenya and I returned to Kansas.

I figured Cynthia would never remember me, so I had a friend take a close-up photo of my right ear, a close-up photo of my left ear, and a head-on shot of both ears pushed forward, and sent them to her. She made an I.D. reference card to carry in her Land Rover file, and I get a Christmas card from her every year.

〰 〰 〰

By 2006, the degenerative nerve condition in my legs had deteriorated to the point that I could walk but a short distance on a flat surface with the aid of a cane or walking stick. Otherwise, I needed to use forearm-cuff-crutches on uneven terrain such as a bush path, or be in a wheelchair for a longer distance such as passing through an airport.

On one of my Kenya safaris in 2006, my drivers rigged a wheelchair for me. It was a bit unsteady and rather low to the ground, but I appreciated their effort.

When we arrived at a favorite lodge after several years' absence, there was jubilation among the male staff as they gathered around our vehicle with the greeting, "Cowabunga, you have been lost for so long!" But when the drivers assisted me into the wheelchair, jubilation turned to misgivings, as they were not aware of my condition.

Imagine sitting in a small wheelchair and looking up at a

circle of concerned faces while listening to the repeated entreaty, "Oh, Cowabunga, Cowabunga! What is wrong? What has happened? Oh, Cowabunga!"

I raised a hand to calm them. Expectantly, they leaned closer.

With a dramatic gesture I pointed to my left leg and said, "This leg is dying." Next, I pointed to my right leg and said, "This leg is dying." Then, I pointed to my crotch and said, "But this leg is still strong."

Peals of laughter erupted, and they excitedly ran throughout the lodge to gather all of the female staff, gently pushing them in a circle around me. "Cowabunga, tell the women what is wrong."

Covering my face with my hands, I shook my head back and forth and said, "Oh, no—I can't tell the ladies." More laughter.

The men made sure that the ladies knew of Cowabunga's condition, and I have been told that the story is often repeated with great flair and embellishment.

ꗃ ꗃ ꗃ

It is comforting to know that regardless of how long I may be absent from Africa, I will always be greeted with, "Cowabunga! You have been lost for so long," for then I will know I have "come home."

ꗃ ꗃ ꗃ

NGORONGORO

Life and Death of a Volcano

epoch Africa
eons ago . . .

towering volcano
spewing
vapors, gases
raging fury . . .

molten lava
erupting
exploding, flinging
trajectory throughout . . .

ebbing, waning
dying
collapsing
within . . .

igneous embers
silence
caldera
finis

—Gary K. Clarke, 2013

The Little Man on the Crater Rim

The Ngorongoro Crater is *beyond* imagination.

An extinct volcanic caldera that stretches ten-by-twelve miles, it is covered in vegetation and surrounded by an escarpment, 2,000 feet high, that descends into a life-filled paradise of micro-habitats with thousands of animals.

If you could spend but one day on the African continent, it should be at the Ngorongoro Crater in northern Tanzania.

One of the great surprises of my life happened at the Crater when I had my brother, Steve, on safari there in the 1970s.

We had experienced a full day on the Crater floor, absorbing all that it had to offer: predator, prey, scavenger; life, death, sunlight, shadow; heat, dust, mud, and dung.

After a precarious drive up the narrow, rugged, switchback trail, we returned to the old Ngorongoro Crater Lodge, perched on the rim. Our cabin, with a commanding view, was at the end of the path.

We freshened up and strolled to the dining room, enthralled by the changing moods of the Crater at end of day. By the time we finished dinner and headed back, night had fallen.

As Steve and I walked and talked in the darkness, we became aware of another presence. A little man was silently walking beside us—from the lodge security staff, accompanying us down the path to our cabin. I use the term "little man" *not* disrespectfully, as he had significant responsibilities for the safety and well-being of lodge guests. In KiSwahili he was an *Askari,* meaning a guard. But, physically, he *was* a small man, slight in stature. Dressed in a well-worn safari uniform, overcoat, cap, and wellington boots, he carried with him only a torch (flashlight), a primitive bow, and a fistful of hand-made arrows.

This was our protection? All kinds of wild creatures roamed the Crater rim at night, some of them the most dangerous in Africa.

Poking Steve with my elbow, I asked the little man, "What would you do . . . if we met a Cape buffalo?" He stopped on the path, poised the bow, arrow in place, then aimed to the left, forward, and to the right, drawing the string a little each time as though he was shooting, and said, each time: "I just go tszh, tszh, tszh." HA!

The little man must have sensed that Steve and I were a bit dubious, as he asked us, "You want to see buffalo?" Together we said, "Sure."

The little man switched on his torch, the beam piercing the night; and there, mere yards away, stood three . . . *Cape buffalo bulls!*

An apparition, like three nocturnal ghosts!

Never before had I experienced Cape buffalo as I did at that moment.

In darkness.

On foot.

So close.

Muzzles wet and shining; nostrils flaring and searching; eyes riveted on us; massive horns curved like twisted sabers.

Such power. Such dignity.

For me, the living expression of the Cape buffalo has more quintessential Africa in it than any other animal I know; more than the roar of the lion; more than the supremacy of the elephant; more than the flight of the fish eagle.

Each time I see a Cape buffalo, it is as though I'm seeing one for the first time, with the same feelings of awe, wonder and fascination.

But *this* time was *so* different, and completely unexpected. It is indelibly imprinted on my mind.

What I found incredulous was that the little man knew all along the buffalo were there, and that the buffalo accepted his presence. These same three Old Dagga Boys grazed this area regularly, and throughout each night the little man and the buffalo knew well each other's movements and location.

Steve and I had witnessed a remarkable rapport—an

understanding and communication between human and animal—a close relationship of mutual trust and respect.

A special gift had been bestowed . . . by three magnificent Cape buffalo and a wonderful little man, on the Crater rim.

த் த் த்

Safari, So Good (as told by . . .)

"Send me a postcard from Africa!"

Prior to each safari, this request was voiced by dozens of people. It was exciting to receive mail from exotic destinations on the African continent, especially for stamp collectors. Many African postage stamps feature excellent animal art work or cultural themes, but it was not easy to dispatch mail from Africa.

So, to simplify the process and save myself time and trouble, I published several postcards featuring photos I had taken on safari, plus a card featuring the Cowabunga Safaris logo.

For each safari I'd bring along a supply of postcards, together with the address list in my safari journal. During the long flight to Africa I would address each card, and in the message section I would write:

Jambo:

Safari, So Good.

Sooo many animals.

Wish you were here.

COWABUNGA!

On arrival in Africa, I'd purchase postage stamps post-haste, mail the cards, and be done with it. That way the postcards usually got back to the USA before I did. One of my safarists used the same procedure—except—he did everything prior to departure, *including* putting USA postage stamps on the cards! That did not work in Africa.

However, there is a far more significant connotation to the phrase *Safari, So Good*.

My fellow safarists frequently conveyed the adventure and joy of being on safari better than I could—through their short stories, odes, poems, jokes, limericks, songs, et al.

It is my pleasure to present, from a smattering of Cowabunga

Safaris Alumni, some of their personal expressions of the safari experience, as well as a cameo commentary from a long-time zoo colleague.

Safari, So Good.

꧰ ꧰ ꧰

There's a Warthog Under My Bed

(... as told by Kay McFarland of Topeka, KS while in Zambia, 1991)

The mascot at Kapani Camp in South Luangwa National Park, Zambia, is a warthog appropriately named Miss Piggy. She is a true bread lover. I had put some leftover bread in my purse after lunch which, with camp staff permission, I dropped on the path for Miss Piggy. She consumed it in short order. When I went to my room a few minutes later, I put my purse, which had held the bread, on the bed and entered the bathroom. Hearing strange sounds, I checked the matter out. Miss Piggy had surreptitiously followed me and was now standing **on** the bed with her head in my purse, looking for more bread!!

I was unfamiliar with the proper protocol to follow in such situations. When I pulled the purse off her head, Miss Piggy jumped off my bed and backed under the other bed, which was against the wall.

I understand now that this backing into a den is warthog SOP. Just her snout was visible. Dragging her out proved impossible. Leaving her there while I went on a game drive did not seem like a good idea.

Having no more bread, I tried laying a candy trail to the door. Miss Piggy checked out the candy but did not care for it. She quickly backed under the bed again. This time, she took hold of the throw rug between the beds and furnished her new den with it. This gave a distinct impression of permanence. It is difficult to order a warthog out of one's room.

About this time Frank Carlson from the next room heard the confusion and came over. Between the two of us we made enough noise that Miss Piggy left in a huff. Later I found out Miss Piggy lived under a bed in one of the camp staff bungalows.

Miss Piggy is a real doll, but definitely piggy about food. She provided a true hands-on-warthog experience.

卍 卍 卍

At the Waterhole

(...as told by Bob Burke of Spokane, WA while in Zimbabwe, 1999.)

They came—out of the shadow they came—huge lumbering-but-graceful giants with that distinctive swinging, rhythmic, ground-eating stride that once seen, is never forgotten. Out of the bush they came in what seemed to be endless columns of gray, almost ghost-like shapes—all sizes—each drawn as if by a magnet to that life-sustaining commodity—water. For it was July, the dry season in the Zimbabwe bush. Rain had not fallen since January and there would be at least four months before the rains would fall again. The only water was at waterholes, scattered

throughout Hwange National Park, and some of them were dry. All creatures, great and small, that inhabit that area of the African bush were making their daily visit to the remaining open holes in search of that precious life-giving water.

There was order in the scene, as there is always order in nature. No one argues with an elephant, so the smaller creatures maintained a respectful distance, waiting their turns. Or, if the waterhole was not too crowded, a giraffe might drink a few feet away from an elephant. But the zebra, the waterbuck, the impala, the wildebeest, even the buffalo, and all the smaller creatures keep a wary eye out for their huge companions—and for the predators that lurk in the shadow waiting for their evening meal. Lion, leopard, hyena, wild dog, jackal—all come to the water, not only to drink but to eat!

⌐⌐ ⌐⌐ ⌐⌐

Safari Limericks

(…as told by Rod Furgason of
Topeka, KS while in Tanzania, 1999.)

There once was a guy named Gary.

Whose hair was so thin it was scary.

So his friend made him grin

From his ears to his chin

By giving him a hat that was hairy.

There was an old boy named Dagga.

Whose puns were so bad they would gag ya.

Though his clients complained

His jokes stayed the same

And today he continues the saga.

There once was a man named Shetani.

Whose job was to go on safari.

One night he got drunk

And stepped on a trunk.

Now the elephants call him hatari!

⊐ ⊐ ⊐

Thanks, Gary

(. . . as told by Jane Michener of
Topeka, KS while in Kenya, 1994.)

At Sweetwaters Camp, to the river we did go,

Where we heard and saw our first hippo.

Then Gary made an interesting find,

And showed us what the elephants left behind.

He even modeled it upon his head,

Called himself a ---- head, is what he said.

Now for the main point of this rhyme,
Thanks for a super, good time!

ꊡ ꊡ ꊡ

A Special Sighting

(...as told by Clayton Freiheit of
Denver, CO while in Tanzania, 1991.)

Game drives on safari are never dull. The objective is to leisurely drive around attempting to locate predators, as they often influence the behavior or visibility of other animal species.

On a Cowabunga Safaris recent safari to Tanzania, we located a mother cheetah and her six-week-old cubs.

We were enjoying the luxury of our private tented camp near Lake Ndutu, or Lagarja, if you prefer, in Tanzania's magnificent Serengeti National Park. The great wildebeest-zebra migration was in full swing and the Ndutu end of the park teemed with wildlife. The early morning game drive had been interesting but not especially

spectacular; only one vehicle went back out after breakfast. They discovered the cheetah family and were the envy of everyone who had stayed in camp. At four p.m., relocating the cheetahs became a high priority!

Luckily, we found them. We noted that one cub was always straggling behind and was also remarkably smaller than his siblings— probably the runt of the litter. I think that his chances for survival are quite slim. We watched for the better part of an hour, being careful not to disturb them. Cheetahs are regularly seen, but unlike lions or hyenas, never promised, due to their nomadic habits and secretive nature. Young cubs are especially attractive with their bushy, gray manes that are later shed when they are about eight months old.

Two days later, as the afternoon game drive departed camp, ours was the last car. As driver Juma held back to allow the dust from the other vehicles to settle, the cheetah family appeared in an open area. We flashed our lights at the others to try to alert them but to no avail. As we approached the mother ran back toward a shady bush to chase off a marabou stork—a sure sign that her kill was concealed there—and the four tumbling cubs followed. They settled down to feed on a young Grant's gazelle for about 45 minutes. Finally mom picked up the prey to move it to a more secure spot while the youngsters frolicked with one another. One even climbed into the low crotch of a small Acacia tree making an excellent series of photos. All this happened within 30 feet of a safari vehicle full of happy safarists. A very special sighting, indeed.

Ɱ Ɱ Ɱ

Cowa-Cowabunga

(... as told by Bob Hassur, Jean Reynolds, Bob Lee, Carol Tantillo of Topeka, KS while in Kenya, 1990.)

(To the tune of Davey—Davey Crockett)

Cowa—Cowa-bunga
A word that our leader lives by
Cowa—Cowa-bunga
For it our leader would die.

The Maasai Mara is where we are
From the U.S. we've traveled far
Our leader has been a Clarke named Gary
Who now is on his 29th safari.

In Nairobi it all began
We got into our Nissan vans
Bumpy roads and dusty air
Gary promised—and they were there.

Old Devil Man has lead the folks
With funny puns and off-color jokes
From camp to camp we have trekked
Is "Cowabunga" really animal sex?

A stick with a knob Gary would carry
Also a dunk stick went with Gary
Mzee Shetani is his name
For fun and wild life he is game.

Because he ran the Topeka Zoo
In the animal world nothing was new
But most of us were virgins in the bush
To daily game drives we were pushed.

The beautiful animals all of us saw
Have filled each heart with wonder and awe
Kundi Bora means we're the best bunch
And none of the animals ate us for lunch.

But we've eaten well, hale and hearty
Except for the few with disease of the potty
We've traveled Kenya from game park to park
And always been glad to have along Clarke.

So now it's all over—the last night dinner
We'll all *go* home feeling like winners
How can we thank Gary for his loving care
A poem and gifts from all who were there.

ᴎ ᴎ ᴎ

Safari

(...as told by Irene Hommert of Kirkwood, MO while in Zimbabwe, 1994.)

A Safari can best be described as a sensory experience.

Awakened to behold the beauty of the first morning light, one catches the scents of dung and wild animal musk.

As the day unfolds, a graceful giraffe is spied as she gently and expertly strips the thorny acacia of its leaves.

Powerful is the elephant searching for food as he uproots a century-old baobab tree. The surge of stampeding wildebeest can be felt as they are seen crossing the savannah. Spine-tingling is the call of the magnificent fish eagle soaring overhead. The lion, devouring a young zebra, gives visual definition to the phrase "circle of life."

Evening approaches.

The fiery orange sun begins to sink into the horizon. Safarists sit by the campfire sipping an indigenous brew. For those with adventurous palates, a sampling of a local delicacy is savored. Conversation is shared and the day's events are reviewed and recorded.

Eventually the sun gives way to the moon.

A galaxy of stars illuminates the evening meal, after which the group makes its way back to the campfire to be lulled to sleep by the dance of its flames.

A permanent fellowship has replaced the politeness of the earlier hour.

Their solidarity is Africa!

꙲ ꙲ ꙲

A Moment

*(. . . as told by Marshall Clark of
Topeka, KS while in Kenya, 1990.)*

Jonathan Leakey's Island Camp on Lake Baringo in Kenya can
get awfully hot in the middle of the day, so most folks stick around
camp to rest or swim in the pool. I didn't. I walked down the path,
past the Njemps village to the lakeshore and found a nice boulder
to sit on while I watched the shore birds and enjoyed the warmth
and the lake.

Quite soon, I realized there was a young boy of perhaps nine or
ten standing 20 or 30 feet away watching me. He was beautifully
black and naked except for a small circlet of beads around his lower
waist. His skin was a smooth and lustrous ebony punctuated only
by huge, limpid, intelligent eyes. He didn't speak, and neither did I.
I picked up a small flat stone and, without looking at him, skipped
it out over the lake . . . six or eight skips. He audibly caught his
breath. I skipped another. The boy moved closer, so I held up the

next stone and turned it back and forth so that he could see its shape before I skipped it. We still didn't speak. Finally, he picked up a stone of the right form and was successful with his first skip. When it was time for me to go, I arose and started up the path. The boy, who by now was standing close to me, took hold of my little finger with his hand and, still silent, led me on the path through his small village and then to my tented camp. We two friends walked, with great pride in our friendship, through the village, hand in hand. Not a word was spoken during our entire friendship, but for a moment in time, two utterly different worlds touched.

That moment will always be with me and, I hope, with him.

Africa!

卍 卍 卍

Safari Personalities

(... as told by ANONYMOUS

while in Kenya, 1980.)

The greatest Safari ever trekked is now done,

The doldrums have come in like the last setting sun.

So, alas and alak,

We can't soon go back

But we can relive some of the fun.

Harry, the cattleman from our west

At napping was really the best

When the rhinos were bleeping

He was still sleeping

His libido has failed the test.

The Rudnicks of Topeka were there
And what a sensational pair!
Their deportment was swell
As far as I could tell
But they sure caused the Maasai to stare.

Janet, Gary's darling daughter
Got sicker than she really oughter
I knew she was on the mend
When she came around the bend
Wearing the new Afro outfit he bought her.

The cry of the red-headed baboon
Sails across the lake like a loon
It's just Gary Clarke
Right after dark
And nothing to cause you to swoon.

Dorothy and June are so nice

Didn't really go in for the spice

But a little smile broke

When Gary told the joke

About the Njemps who could only jump twice.

Then there's the fractured-wing Zoolie

Who bore up with that arm like a coolie

But when it came time for a drink

One could only think

The whole thing was a great big foolie.

Lois and Suzanne bunked together

And stood up against all the weather

But when all was said and done

Shopping frenzies were really the fun

Except that you can't take home any leather.

From Turkana, Maralal and Narok

Brenda's pen never missed a stroke

Through the bumping and whirling

And the red dust a-swirling

It's a wonder her writer ain't broke.

Georgia Sue, Georgia Sue, Georgia Sue

What can we say about you?

You're most loveable of all

But you sure had the gall

Serving up your zesty conversational stew.

And now for the last of the troupe

Whose eyelids never did droop

He saw every view

Even made up a few

And fell in love with the whole gall-danged group.

There's a Lion in the Gents!

(. . . as told by Clayton Freiheit of
Denver, CO while in Tanzania, 1998.)

Last February I was delighted to make my tenth visit to the East African nations of Tanzania and Kenya.

As our small charter plane proceeded northwest toward the great Serengeti National Park in Tanzania, we overflew both Lake Manyara and the Ngorongoro Crater. Over the Serengeti, we could see the many kopjes (kop-ees), the characteristic outcroppings of granite rock emerging from the vast sea of grass which extended from horizon to horizon. Our destination was Seronera, which serves as the park's headquarters and is also home to the Serengeti Research Institute as well as Seronera Safari Lodge. Prior to landing, our pilot had to buzz the mowed grass airstrip to chase off small herds of impala and gazelles that were grazing there.

Seronera Safari Lodge, built in the late 1960s, is imaginatively designed and sited around a large kopje so as not to intrude on the natural landscape. It is unfenced and wildlife abounds in the area due to the nearby Seronera River which offers water year-round. Rock hyrax, agama lizards and baboons are at home at the Lodge and several years ago, I drew drapes in my room to find an adult cheetah standing outside the window only a few yards away!

The Lodge's bar and dining area is handsomely integrated into the buildings; it is reached by a series of wooden stairs and landings and steps hewn into the rock. This is African lodge ambiance at its best!

While our group was having dinner, a spectacular thunderstorm occurred which lasted for over an hour and made an after-dinner drink in the nearby lounge with its wood-burning fireplace seem like a fine idea. By 10:30, John Baird, Gary Lee and I were the only customers left and we decided to turn in.

As we descended the stairs and approached the last landing overlooking the reception area, we could see a number of the Lodge Staff peering anxiously below. We were soon warned, "Don't go any further Bwana, there's a lion in the Gents." Of course we had to

see this, so John, Gary and I went to the front row to observe. Sure enough, a few minutes later, out of the doorway of the men's room strolled a full adult lioness! She looked up at the crowd and was obviously both confused and angry as she hurried through the Lodge lobby and out the front door. What a fantastic T.A.B. (That's Africa Baby) and great adrenaline rush, something I hope for on every Safari.

As the three of us carefully walked back to our rooms, we were both amused and thrilled by the experience. Over breakfast the next morning, as we recounted our adventure to the rest of the group, the waiter gave us the sequel to the story. After she fled out of the reception area, the lioness circled around the back side of the kopje and climbed up the outcropping where she entered one of the bar's open doors and strolled through the lounge before exiting through another portal. We had another adrenaline surge as we speculated what might have happened the night before if we had decided to have another drink. What was the lion doing? My guess is that she was searching for a dry place to lie up for the night after the severe rain storm, and our paths happened to

cross. Memories such as this are what good safaris are made of! COWABUNGA!

꜀ ꜀ ꜀

An Ode to Gary Clarke (The Mzee Shetani Sonnet)

(…as told by the Thiele family of
NE, AZ and KS while in Kenya, 1995.)

There once was a man named Gary,

The top of his head was not hairy;

He spoke with a grin,

Armed with torch and pen,

And hassled both Larry and Kerry.

From the Sun Belt and Coeur d'Alene,

Topeka and Omaha out in the plain;

From Texas to Missouri,

They came in a hurry,

To catch that Johannesburg plane.

Clear instructions Mzee Shetani did bark,

During the morning, noon and at dark;

He would look for and find us,

And constantly remind us,

That we are not at a theme park.

He led us like a kind army colonel,

And pleaded to keep up with our journal;

He mimicked animal sounds,

He kept much laughter around,

And his sleeping habits showed he's nocturnal.

Oh, the family safari—first rate,

As for Cowabunga group, "You're doing great;"

On game drives they went,

Video by Larry and Kent,

When Kent's diet was not Kaopectate.

The mammals we saw herds and herds,

It compensated for those lousy birds;

Mama Tembo drank sherry,

She competed with Gary,

And the puns they shared were absurd.

At Sweetwater we saw the giraffes,

Drinking water they caused many laughs;

Samburu gave many a thrill

With the cheetah chase and kill,

But those damn Yankees kept crossing our paths.

A highlight of the trip was the Ark,

Watching animals come out from the dark;

Took a plane to the West,

The Mara has been the best,

For those mammal watchers like Mark.

Seventeen lions in a pride,

Were seen crossing a pasture to hide;

Time to quench their thirst,

When Gary said "This is a first".

Then we heard Larry say,

"Oh shit, I don't have my camcorder today!"

And the Maasai, they were quite willing,

For the price of a few measly shilling;

To chant and to dance,

Gary gave us the chance,

Mzee Shetani lived up to his billing.

With Mama Makora as first mate,

Gary kept all things running straight;

The Tusker did flow,

It made him choo and choo,

But kept him at his ideal safari weight.

Tomorrow we leave, we're not staying,

Gary would like to, that goes without saying;

But grandkids and wife,

Makes it a double life,

For the man whose African band won't stop playing!

We hope this poem gave you a smirk,

Having you as our guide was a perk;

We leave here tomorrow,

Our hearts filled with sorrow,

For now we must go back to work!

T.A.B

(That's Africa, Baby)

꙰ ꙰ ꙰

Thanks for the Memories

(...as told by Bob Burke of
Spokane, WA after numerous safaris.)

(to the tune of Bob Hope's signature song)

Thanks for the memories
Of candlelight and wine
Of hippos where we dined
Of dusty miles and crocodiles
Who tried to grab our line.
Yes, thank you so much.

And thanks for the memories
Of leopards in a tree,
Of baboons on a spree,
Of elephants who stood on guard

So we could drink our tea.
Yes, thank you so much.

And, thanks for the memories
Of vultures perched in trees,
Of warthogs on their knees;
A hyena's feast, and wildebeest
As far as we could see.
Yes, thank you so much.

So thanks for the memories
Of lions in the night
They caused us such a fright.
Then our gaze we tossed to the Southern Cross
It was a gorgeous sight.
So thank you so much.

囝 囝 囝

April 2, 1996

Mr. Gary Clarke
Cowabunga Safaris (Pty.) Ltd.
Private Bag 4863
Gage Centre Station
Topeka, Kansas 66604-0863

Dear Gary:

Following you through Tanzania was disheartening. Everywhere we went, the animals complained that they were all looked out. Many had personal identification problems.

Three impalas told us (separately) that the Cowabunga Crowd had passed them off as elands while a zebra complained that he had been identified as a Topeka tourist.

Naturally, it is painful to have to bring these problems to your attention (I shall say nothing of the ostrich who worried that he had almost made the mistake of incubating your head) but that is what old friends are for.

All best,

William Conway

How I envied these Old Dagga Boys as they soaked under the snow and cloud-shrouded magnificence of Mt. Kilimanjaro.

Amboseli/Kenya

I <u>see</u> the commanding presence of regal lions; I <u>feel</u> the heat as it radiates from the kopje rock; I <u>hear</u> the wind as it touches the King's mane; I <u>sense</u> the immensity and splendor of the African wilderness.

Serengeti/Tanzania

These giraffes appeared out of a torrential downpour as if by magic . . .

Maasai Mara/Kenya

My brother discovers the charm, the intrique, the romance of Africa . . .

Stone Town/Zanzibar

A hyena filching a hunk of carcass . . . death is an element of life on the African savannah.

Serengeti/Tanzania

The monkey's tale.

Samburu/Kenya

Visiting with these cashew nut merchants was uplifting and inspiring. I admired their sincere demeanor, as well as their unpretentious shop—so refreshing in a phony and convoluted world.

Dar es Salaam/Tanzania

THE SKULL OF
AUSTRALOPITHECUS BOISEI
(ZINJANTHROPUS)
WAS FOUND HERE BY
M. D. LEAKEY
JULY 17TH, 1959

Our guide stands on the "floor" of Africa.

Olduvai Gorge/Tanzania

Misadventures
Out of Africa

How Are You Going to Get That Home?

I was basically a non-shopper on safari. If anything, I would haunt bookshops, both old and new, for out-of-print items or books published in Africa that were not available in the USA.

Members of my Cowabunga Safaris groups, however, usually returned home with conventional mementos: small carvings, traditional jewelry, local handicrafts, tee shirts and safari-style clothing—often purchased at a roadside duka, which is KiSwahili for 'curio shop.'

Invariably, there were those who could not resist a gigantic carving or collectible for their den, patio, or newly-conceived "African room." I would always shake my head in dismay and chide them, asking, "How are you going to get that home?"

And then one day on the edge of the Great Rift Valley in Kenya, at a duka I had visited many times, I was *awestruck!*

Never . . . in my many years all over Africa . . . had I seen such a carving. It was stunning: a life-sized depiction of a young gerenuk in the distinctive upright feeding posture! EXTRAORDINARY!

The gerenuk, pronounced 'ger-a-nuk' with stress on the first syllable, is an elegant, thin-legged, long-necked antelope, found in the semi-arid bush of East Africa. Only the males have horns. Almost exclusively a tree-foliage browser, the gerenuk habitually rises on its hind legs to reach vegetation more than six feet up. It is sometimes called the giraffe gazelle. (See the photo on page 25 of my book, *Gary Clarke's AFRICA*.)

While my group shopped throughout the duka, I stood entranced, admiring this lovely carving. It had cast a spell on me. The manager noticed, walked to my side, stood by me in silence, and said reverently, "Ah, Cowabunga; a beauty, eh? I have never seen anything like it."

"Me neither," I replied.

Then the manager started his sales pitch. "Ah, Cowabunga; this is a one-of-a-kind piece of wildlife art. What do you offer for it?" Dukas operate on a barter system. The vendor starts high, you start low, and hopefully both parties agree on a price close to the middle.

I said, "What do you ask?"

He said, "$500.00."

I said, "$100.00."

He said, "$400.00."

I said, "$100.00."

Reluctantly, he said, "$300.00."

I said, "$100.00."

After a silence, he said, "OK; $250.00, best price."

Even though I thought it was well worth the $250.00, I said, "$100.00 my friend, and thank you. I respect your fair price, but I really don't need this."

Hesitating, he finally said, "$200.00, final offer."

I felt bad, as I was *not* trying to beguile the man. It truly was a lovely carving, but how would I get it home?

So, to end the transaction, I said, "$100.00, and I know you cannot accept that, but I must go, as my group is waiting for me. Thank you."

As I was about to get in the safari vehicle, the manager came running after me, calling, "Ah, Cowabunga, Cowabunga; you must have the gerenuk—OK, $100.00; it's yours!"

I was overjoyed, as I knew just the place for it back at Cowabunga Safaris Main Camp in Topeka.

But my euphoria was quickly overshadowed by merciless teasing and verbal abuse from my group. I had to swallow my pride as a non-shopper in order to endure, over and over again, from each individual, the dreaded question, asked in a blatant manner: "HOW-ARE-YOU-GOING-TO-GET-THAT-HOME?" It was all in fun and I deserved it.

I must admit that I was in a conundrum, but I had a plan.

On our return to Nairobi, I arranged to get some heavy-duty corrugated cardboard and fashioned it into a long oval tube, somewhat like a golf bag. The carved gerenuk, wrapped protectively in newspaper, fit perfectly. The tube was secured with end panels taped to the top and bottom. Tying sisal rope around the tube, I made a shoulder strap so I could carry it aboard the plane (this was long before today's stringent airline regulations). Once on board, I told the flight attendants about my rare and fragile work of art, and asked if they had a safe place to stow it. On each flight segment—both international and domestic—the cabin crews were most accommodating. I was greatly relieved that I never had to check my precious cargo.

And so today, by my desk at Cowabunga Safaris Main Camp, a gerenuk stands tall, nibbling on the green leaves of a live plant, reminiscent of an unforeseen shopping saga.

卍 卍 卍

African Killer Bees? Preposterous!

When visitors walk through the entry to Cowabunga Safaris Main Camp, they pass under a rather substantial log suspended by sisal rope.

Looking up, they ask, "What is it?"

I reply, "Obviously, it is a melliferous *mzinga*."

They respond, "Well, ok, but what *is* it?"

Melliferous is from the Latin *mel* (honey) and *fer* (bearing). *Mzinga* is KiSwahili for beehive or honey trap. Here's how it works.

Indigenous Africans will take a sizeable log and split it lengthways in a three-fourths/one-fourth ratio. The large section is hollowed out and becomes the top of the *mzinga*. It will be positioned so that the hollow area faces the ground. The solid, small section serves as the bottom and is secured to the top part with sisal rope or wire. It is not a tight fit, purposely leaving a narrow slit the length of the log between the large top section and the small bottom panel.

The *mzinga* is then placed in the crotch of a tree or hung from a horizontal limb. With access through the narrow crack, honey bees will establish a hive and store their honey. Periodically local villagers smoke the bees out of the *mzinga*, open it and remove some (but not all) of the honey, and then place it back as it was.

In my travels throughout Africa, I had seen numerous *mzingas* of various shapes and sizes. They are particularly prevalent in the Ardai plains of northern Tanzania.

I had always considered them to be a fascinating cultural artifact—one worthy of an honored place in Cowabunga Safaris Main Camp. I casually mentioned this to several individuals during one of my safaris, and it seems that the word was out in the bush that "Cowabunga needs an *mzinga*." Sure enough, one showed up in Arusha Town with a tag that simply read: "For Cowabunga."

It was a beauty, measuring nearly three feet in length, thirty inches in circumference, and weighing—a lot! Such a splendid *mzinga,* and genuine! Fortunately, it had not been used for some time. This was to my advantage, since there was no indication of bees or honey. Apparently it had been kept in an undisturbed place, because when I opened it dozens of small lizards erupted out of the cavity.

Due to its size and weight, it was too cumbersome and heavy to carry on the plane. I knew that I would have to check it as special luggage. No problem: it was solid and sturdy enough to withstand rough handling.

My only concern was how to clear my *mzinga* through customs into the United States. I felt that the agents probably would not know what it was, as even I had never seen an *mzinga* outside of Africa, not even in a museum. If I said that it was a beehive or honey trap, the reaction might well be, "African Killer Bees! You can't bring *that* in!"

If so, I envisioned responding, in my affected British accent, with a haughty, "African Killer Bees? Preposterous!" But then, I thought not.

How could I explain this wonderful acquisition?

What would make sense?

Would it seem plausible?

On my flights during the long journey back, I mulled over these questions. All of my documentation would show I was coming from Africa, so I figured the *mzinga* should somehow relate to either African art or culture.

Upon arrival at JFK, I had to claim it at the "Special Luggage Area" and then proceed through customs. It seemed to catch everyone's attention and raised a great deal of interest. The moment of truth was upon me. Hopefully my explanation would pass the ultimate test.

So, as the curious custom agents gathered around my prized *mzinga*, I turned it upside down, removed the smaller bottom panel like it was the lid, pointed to the hollow chamber in the large, lower section, and said: "This is a tribal elders' ceremonial beer log that is used for the fermentation of traditional brew."

It worked. Cheers.

 ☢ ☢ ☢

Custom-Made by an Elephant

Once upon a time in Africa, not far from the fabled Zambezi River and upstream from Victoria Falls, there was an "elephant camp" devoted to the guardianship and nurturing of orphaned elephants.

I had the great joy of staying at this camp, and participating in the daily activities of these foundlings—their feeding, bathing, grooming, playing, socializing, foraging, resting, as well as their trail walks through the bush. Some of the elephants were quite young and small in stature, and you walked alongside them on the trail. Others were older, and large enough to ride.

Within several days a conditional bond developed between the elephants and me. I had learned the elephants' personalities and individual behavior quirks. In turn, they knew my distinctive scent, voice, and specific mannerisms. What an honor, several times a day, to be greeted by curious elephant trunks, dripping with mucus, snaking over my face, head and neck, probing my shirt, my shorts, even my shoes . . . and then to rub the elephants under the eye, or to scratch their tongue, and be rewarded with a deep rumble of contentment.

During one trail walk I was privileged to ride on the largest elephant, a bull named Jock. I had ridden elephants before, and always felt like an innocuous nodule on the largeness of the world's most imposing land mammal.

I sat astride Jock's back just behind his groom. The groom, a young African man, would theoretically guide Jock on the trail and, with voice commands or cues, instruct Jock to move forward, to stop, and so forth. Out on trail, the warm sun, the gentle breeze, the chorus of bird calls, and Jock's rhythmic movements lulled me into a comfortable, relaxed state.

Unexpectedly—Jock bolted off the trail and made a dash toward a large tree in full leaf!

The groom urged Jock back to the trail, but to no avail. Jock forged ahead to the tree, and before we knew it, the lower limbs and branches and leaves were hitting me and the groom in the face, on the shoulders, the chest and belly. We threw up our arms for protection, twisting and dodging as best we could, trying not to get knocked off.

Jock stopped at the base of the tree and lifted his massive head. Then I saw his powerful trunk, with over 40,000 muscles, snaking and winding up the tree. I watched his trunk wrap firmly around a large limb, and I witnessed Jock *break it completely* off the tree with a loud "<u>CRACK</u>."

AWESOME!

Carrying the limb, with its profusion of luxuriant green leaves in his trunk, Jock proceeded calmly down the trail as if the entire incident had been the most routine of activities. Of course, for an elephant, it was. So often I had seen other elephants demonstrate this behavior, but this time, I was part and parcel of the entire experience. What a thrill!

But, the best was yet to come. Having the advantage of literally being in Jock's world from my high perch, I was able to observe the next chain of events. It was fascinating.

Now that Jock had the leaf-filled limb at hand, or maybe I

should say "in trunk," his entire demeanor changed. He seemed oblivious to his surroundings and was totally absorbed in the limb.

While holding it in his mouth, Jock grasped leaves with the two fingers of his trunk, and then—with uncanny dexterity—somehow placed the leaves in his mouth, holding the limb in his trunk while he chewed! Once all the leaves were stripped off and consumed, Jock concentrated on the bark. Grasping the limb with his trunk, he manipulated it back-and-forth in his mouth—like a corncob—and chewed off the bark from one end to the other—almost. The narrower end retained a small patch.

Next, Jock did something that was amazing and profound: he lifted the limb high in the air, then back over his head—and presented it to me! It was wet and slick and glistening with elephantine salivation, but *it was wonderful*, and quite an emotional moment for me.

Whatever this gesture may have meant to Jock, it provided me with a tangible memento of an incredible encounter with an extraordinary elephant.

It does seem fitting to have a proper, ceremonial staff—custom-made for me in Africa by an elephant—that I parlay with a flourish, on grand occasions, to exude my rightful stature as the President-for-Life of Cowabunga Safaris.

And what fun it is to recount the remarkable tale of how it came to be.

೩ ೩ ೩

The Cane That Kept Its Secret

The wizened Chief was weathered and pockmarked, reflecting his many years of life in a harsh desert environment. He had befriended me on my previous safaris to the remote and hostile Northern Frontier District of Kenya. Now we met again, and he welcomed me with a gift.

It was a cane.

But it was not of the traditional carved-wood design. Rather, it was a cane that had been fashioned from a hollow metal tube and covered in camel hide, featuring an artistic hand-tooled design along its entire length, from the metallic tip to the curvature of the handle.

Holding the cane reverently in his hands, the venerable Chief leaned toward me, and in a solemn voice said, "Cowabunga, this is not an ordinary cane. This cane is powerful and commanding. This cane will keep its secret no matter where it travels."

What did he mean?

Was it something to do with a tribal belief? A special ceremony? An ancient ritual?

Then, gripping the elongated shaft in his left hand, he pulled firmly on the handle . . . and withdrew the blade of an imposing sword!

NOW I understood!

Under the guise of a common implement used for assistance in walking, this cane could *indeed* keep its secret of being a formidable weapon as well. And it seemed quite normal for an *Mzee* (KiSwahili for Elder) like me—bald head, full beard, pot belly—to use such a cane.

Back in Nairobi, packing for the international flights, I was faced with the quandary of how to transport this sword-hidden-in-a-cane from East Africa to Kansas. It was too long to fit in my carry-on bag. I had no checked luggage. I was reluctant to check it as an individual piece, even if packaged in a box or bag. So I decided on a procedure I had used successfully in the past for spears and knives (this was pre-9/11).

I tied a Cowabunga Safaris luggage tag (with my name) on the handle and then used it as a standard cane all the way to the airport security check-point. *This is what I expected next*: Once the cane went through the x-ray and the sword blade was evident, I would be told I could not carry it on board. At that point I would ask to see a flight crew member and request that *they* carry it aboard, store it in a safe place, and return it to me at the end of the flight. This was a routine practice in those days, and flight crews were most obliging!

I placed the cane on the conveyor belt, proceeded through the metal detector, and waited on the other side for the cane. It emerged from the x-ray machine, the security guard handed it over, and waved me on!

I was so surprised that I continued to the departure gate and then on board the aircraft, fully expecting to be detained at any moment.

But . . . no.

I found my assigned row, carefully placed the cane in the overhead compartment, sat down and fastened my seat belt.

HA! I figured the equipment at the Nairobi airport may have been below par, but I knew that for my connecting flight in Amsterdam it would be high-tech.

I was gobsmacked when the cane cleared security at the Amsterdam airport unimpeded. So I carried it with me on the flight to Chicago, knowing for sure that their equipment would detect it.

But . . . no.

So I flew to Kansas City with the cane in the overhead compartment, and then drove to Cowabunga Safaris Main Camp.

This entire episode was purely coincidental. My intention had never been to violate security with a cloak-and-dagger cane.

Yet it happened.

In retrospect, the words of the Chief concerning this cane—and the aura and mystique associated with it—now seemed prophetic. It truly *was* the cane that kept its secret . . . throughout the entire journey of its long passage out of Africa.

፰ ፰ ፰

Out of the Okavango with the Outlandish

On the Zambezi River, you need a paddle for your canoe.

In the Okavango Delta, you need a pole for your mokoro.

Unlike the stereotypical swamp with murky, still water that harbors death, the Okavango Delta is crystal-clear—dancing and moving— vibrant with life. Each year, fresh rainfall from Angola flows slowly southward and crosses the border into Botswana to create the Okavango Delta. It takes the Angola water several months to reach the Delta, which swells to its capacity in June and July. It is the world's largest inland Delta, yet the waters dissolve in the sands of the Kalahari without ever reaching the ocean.

The traditional means of transport in this aquatic wonderland is by way of mokoro, a narrow dugout made from a hollowed tree trunk. In the past it was frequently an African ebony or a sausage tree. Today, a mokoro is fashioned more conventionally from fiberglass in order to save large trees.

The mokoro is propelled slowly by a long, narrow pole ending with a two-pronged fork at one end. The 'poler' stands in the back of the mokoro, catching underwater vegetation in the fork to push forward.

In the 1980's our access to the Okavango was through a village called Jedibe, on the edge of the Delta. It had a population of 300 and dates back to the 1890s. It had an airstrip which could just barely accommodate our nine passenger Islander aircraft. On the days when a plane landed, it was a big event in the village. Most of the residents showed up at the airstrip, especially the children. Our group required *two* aircraft, so it was a really big deal! After disembarking, we walked through the village and boarded two power boats, which transferred us to our lovely tent camp in a secluded lagoon. The camp served as the assembly point for our overnight excursions via mokoro into the Delta, where we camped on islands with minimum gear.

The mokoro experience is one of the most spiritual in all of Africa, or indeed, anywhere. There is very little freeboard between the dugout and the waters of the Okavango, and you sit below water level as you are poled with a gentle rocking motion through the reeds and papyrus, caressed by the warm sun and the cool breeze.

It is so Q-U-I-E-T. It is one of those times on safari when you do not have to speak, when silence is not unsociable. The human voice would be an intrusion on the Spirit of Africa.

Walking in the bush, you commune with the earth. On a hot air balloon flight over the savannah, you commune with the wind. In a mokoro on the Okavango, you commune with the water.

Most mokoro poles are made from the silver terminalia, a slender tree with hard wood that grows in the sandy soil of the Okavango. Young specimens are favored because of their straightness and resistance to termites.

To the owner, a mokoro pole is like a set of car keys—very personal. The owner knows the feel, the weight, the balance, as well as the responsiveness of the pole in the water.

After our Delta experience, we returned to the airstrip at Jedibe village for our flight on to Victoria Falls. Among the villagers gathered at the airstrip was a man with a mokoro pole! His name was Neashi and we visited. The pole was such an intriguing cultural artifact that I negotiated with Neashi and purchased it for 20 Botswana pula ($10.00). I had never seen one outside of Africa, not even in a museum. I understood why after the convoluted labyrinth that I had to navigate in order to get it back to the USA!

It was a beauty! And—it was 12 feet long! Yes, T-W-E-L-V-E.

My group and the bush pilot were most understanding and helpful. The villagers were entranced as we struggled to fit the pole through the rear cargo door, then carefully maneuver it forward along the entire interior of the aircraft—a tight squeeze and also a bit of an inconvenience for passengers on that side.

From Jedibe, we flew to the border town of Kasane to officially exit Botswana, then on to Victoria Falls in Zimbabwe, where the pole resided for a few days in storage at the classic old

Victoria Falls Hotel. By now it was known as the "Old Dagga Boy Mokoro Pole."

On our return to the Victoria Falls airport, the baggage attendants put a special tag on the pole and hand-carried it with pride to the cargo hold of our Air Zimbabwe four-engine jet, which took us to Hwange National Park. At Hwange, the guides secured it to a Land Rover for travel to and from our Bush Camp. Then I had a bit of a problem.

From Hwange we flew Air Zimbabwe again, with the pole in the cargo hold, to the small airport at Kariba, so we could connect on charter flights to two remote destinations—Bumi Hills on the edge of Lake Kariba, and Mana Pools along the banks of the Zambezi River—in single engine, five-passenger planes. There was no way I could take a 12-foot pole!

The baggage attendant at Kariba airport knew me, saw my dilemma and offered a solution. "Oh, Cowabunga," he said. "I'll keep your pole here safe with me, until you return for your flight on the big plane to Harare." And he did. Such a friend.

Everyone along the way had been fascinated with my mokoro

pole and, despite the length, took great care to see that it was handled carefully. Now, however, I faced the ultimate challenge: to get the pole out of Africa on international flights, with airlines and people who would not know and probably would not care what it was, through a system not designed for such a unique item. And once it started the journey, I would not see it again until I arrived in the USA.

I helped my friend at Kariba airport with the airline routings, airport codes, and extra Cowabunga Safaris luggage tags. From Kariba it went on Air Zimbabwe to Harare, connecting on to London. In London it would transfer to a TWA flight to St. Louis, where I would clear it through customs, then on to the final TWA leg to Kansas City.

On arrival in St. Louis, I anxiously waited at the oversized baggage area . . . and waited . . . and finally, there it was, in perfect condition. What a relief! Now it was just the short flight from St. Louis to Kansas City and then the drive to Topeka.

I proceeded to the domestic ticket counter, obtained my boarding pass, and presented the pole to be checked as baggage.

"What is it?" the agent asked. I explained about Botswana and the Okavango Delta, and about a mokoro and a mokoro pole, and about the long trip out of Africa with all of the special handling, and now almost home. With that, the agent put a tag on the pole and threw it on the moving conveyor belt.

"NO," I yelled, "WAIT! STOP!"

The pole went around a corner with a sickening **SNAP!**

My heart sank.

When I retrieved it in Kansas City I saw that it was badly bent, but still in one piece. The next day in Topeka I took it to a woodworking craftsman. He said that fortunately the damaged area was on the lower portion of the pole, which meant it had been in water much of the time. He called it a 'greenstick fracture' and, since it was fresh, he could repair it.

As I write this at Cowabunga Safaris Main Camp, the mokoro pole hangs over my desk and runs nearly half the length of the Camp. Unless I point it out, you would never notice the fracture. This pole is just one of the many wild and wonderful things I brought out of Africa in the "old days." Those days were not *that* long ago, but the world has since forever changed, particularly with reference to international air travel and what you can transport on an airplane. I could not do today what I did then. Many of the unusual and original items from Africa that once were commonplace are now museum pieces.

卍 卍 卍

Snail Mail from Village and Bush

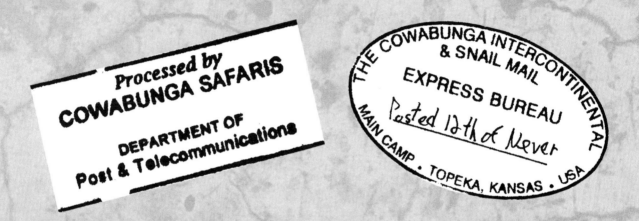

It comes in all shapes and sizes: envelopes that are small, square and white; envelopes that are rectangular and pink, edged with red and blue *Air Mail* bars; or envelopes that are oversized and brown with *Par Avion* stickers on them. Or, colorful post cards of fauna and flora.

Snail mail from Africa is the highlight of any day at Cowabunga Safaris Main Camp in Topeka.

It is not just the excitement of a heartfelt message from treasured friends far away. The envelope is just as important—the hand-written address reflecting humanity that makes the recipient feel special; the colorful postage stamps featuring African wildlife or culture; and the stamp cancellations from exotic locales, sometimes in local languages.

The most unique post, although now quite rare, is the Aerogramme—a single sheet designed to fold up with the message inside. The destination, address and stamps go on the front outside panel, with the return address on the reverse.

I'm not an email guy, much to the consternation of my friends across the globe. While I can usually receive them on my antiquated equipment, I'm unable to answer due to my lack of computer literacy. So, I respond by snail mail. If an immediate email reply is necessary, I telephone my associate Brian Hesse (Mzungu Mrefu) and he sorts it out. I realize that this idiosyncratic quirk annoys some of my correspondents, but they tolerate it for the sake of our friendship, and for that I am grateful.

9-2-07

We miss you and Clayton
but we're thinking a toast
to you every day (it's hard)
Saw your book on sale at
Oldwai Gorge!!! Unbelievable!
Take care and say hi to the
Great Danes for us! Bill

KENYA 95/-

NAIROBI

TANZANIA SAFARI

M ZEE SHETANI
1ST FL. Wgasa House
Dung Beetle Square
Corner Midden Ave + Long Drop
Way

Zimbabwe 2c
HELMETED GUINEAFOWL

Zimbabwe 45c
Zimbabwe

BY AIRMAIL
PAR AVION

19 NOV 13

C.L. Clark
Cowabunga Safaris,
P. Bag: H 863,
Gage Centre Stn.,
Topeka,
Kansas 66604-0
USA

Merry Christmas

It would seem that the value of a postage stamp in most African countries is equivalent to the smallest denomination of the local currency. Granted, most of the mail is a single, lightweight piece requiring minimal postage. In my case, however, I often receive journals, maps, or other large packets from Africa, and instead of one or two stamps of higher value in one corner, the face of the envelope will have numerous stamps of minimal value—as many as 30 or more covering over three-quarters of the envelope surface. I love it—such a T.A.B!

Social interaction is a significant aspect of African culture. If I had not seen an African I had met 10 or 15 years ago, or saw them again in a different locale, their first question was, "Remember me, Cowabunga?"

Then, with great enthusiasm, they would proudly recount when and where we first met, making sure everyone within ear shot knew of our long acquaintance. Some individuals I would see periodically, and each time it would prompt a litany of all of our previous meetings, in a dramatic and detailed fashion, listing dates and places and shared experiences. Such heartfelt

expressions of friendship were always so gratifying to me.

This virtue was also expressed in snail mail letters from village and bush. One of my favorites was from a young lady in Kenya, with the given name of Rollex. Following are excerpts from a lengthy letter, in her own hand:

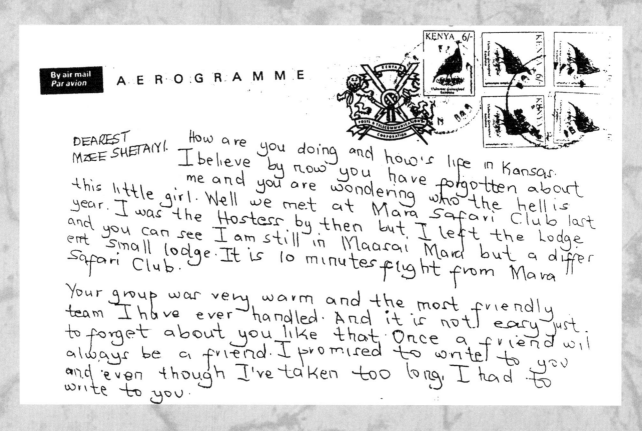

DEAREST MZEE SHETAIYI. How are you doing and how's life in Kansas. I believe by now you have forgotten about me and you are wondering who the hell is this little girl. Well we met at Mara Safari Club last year. I was the Hostess by then but I left the Lodge and you can see I am still in Maasai Mara but a different small lodge. It is 10 minutes flight from Mara Safari Club.

Your group was very warm and the most friendly team I have ever handled. And it is not easy just to forget about you like that. Once a friend will always be a friend. I promised to write to you and even though I've taken too long, I had to write to you.

I took lots of photographs with your group and I wonder
if you developed/printed the photos. I would kindly love
to have a look. You could post them on the address
on the back. I promised my dad your photo. Remember
you were in the Mara in August last. If you have
forgotten about you me. I am a tall slim black chocolate
beauty. Otherwise and I beg to remain and please
keep on smiling and be happy always. Pass my love to
all.
Naughty Girl like Mzee Shetani
Rollex

Snail mail from Africa may take a while to get to the USA,
but I'll give the United States Postal Service credit, in that it
does get here, even with next-to-nothing for an address. As a test
case, I once mailed a post card from South Africa with just the
word COWABUNGA! And my zip code—and it made it!

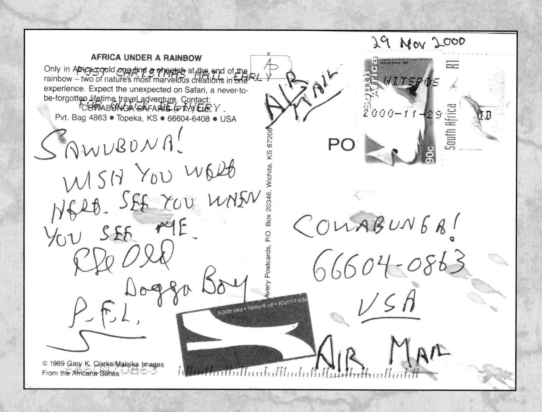

AFRICA UNDER A RAINBOW

Only in Africa could one find a cheetah at the end of the rainbow – two of nature's most marvelous creations in one experience. Expect the unexpected on Safari, a never-to-be-forgotten lifetime travel adventure. Contact:
COWABUNGA SAFARIS (PTY) LTD .
Pvt. Bag 4863 • Topeka, KS • 66604-6408 • USA

© 1989 Gary K. Clarke/Malaika Images
From the Africana Series

[Handwritten:]
SAWUBONA!
WISH YOU WERE
HERE. SEE YOU WHEN
YOU SEE ME.
RBC OLD
Dagga Boy
P.F.L.

[Handwritten address:]
COWABUNGA!
66604-0863
USA
AIR MAIL

[Postmark:] 29 Nov 2000
WITSPOS
2000-11-29
South Africa
R1
90c

Many times on safari, a letter would be waiting for me at a camp with no stamps, but in the upper right corner this hand written phrase:

Safari Mail
By Kind Hand

Of all the official services or systems world-wide that deliver letters or parcels, the above is by far my favorite.

And, truth be told, it is *not* official. It is as unofficial as can be, and totally dependent on the camaraderie and hospitality of anyone and everyone in the African bush, where we are all dependent on the good will of one other.

On the safari trail, where there is no formal or organized postal system, I would ask individuals in transit to other camps or remote destinations to hand-carry letters from me to friends or colleagues I knew were, or were scheduled to be, there. Safari mail was sent to me in the same fashion.

I used it regularly during my many years on safari. What a thrill it was to arrive at some remote outpost or out-of-the way camp and have *Safari Mail* waiting for me. Though sometimes months old, it was always welcome, and never out of date.

Even in the USA I receive *Safari Mail*, hand-carried from Africa by people—often complete strangers—who, while on safari, met someone who knew me and wished to send a letter.

As I page through my extensive files of snail mail and *Safari*

Mail, I'm filled with memories of friendships which would fill a book. Due to space limitations, I'll share just a few with you.

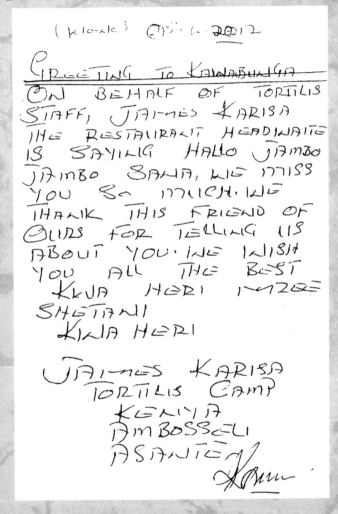

RUFIJI RIVER CAMP
1/7/10

Howzit Ol' Dagga Boy,

Quick note from the banks of the Rufiji River. Your friends Steven & Marjet were here for a couple of days on their safari & they said they knew you! Kansers! How could they not! Even better they have shared African skies

5/14/07

MUZEE SHETANI

WE STILL REMEMBER
You, WILL YOU BE
BACK TO TARANGIRE
SAF LODGE? WHEN!!
WE LOVE YOU COWABUNGA

STAFF FROM
T.S.L
TANZANIA.

Rukoeli,

Dear Old daaga boy

Hope you are well, we
miss you and we all love
you

Your lovely friend

Bonang @ Kwara Camp
Kwando Safaris

I used to be with Okavango
wilderness Safari

Cawabunga Sawa Sawa

In closing this chapter, I'll sign off with a phrase often used by my African friends:

See you when you see me.

౫ ౫ ౫

Translate, Decipher, Interpret (if you can)

In case you cannot read my handwritten statement above, what it says is:

> My handwriting has always
> been bad, and as I get older,
> it is even worse.

All my life, *everyone* has bellyached about it!

My teachers, from primary school to university; my colleagues in zoo biz; my family and friends; and my snail mail correspondents in Africa, from bush guides to bush pilots. They are replete with snide remarks, a few as follows:

"Your recent letter has been submitted to skilled hieroglyphic specialists in Egypt; they are dumbfounded."

"Hey Cowabunga—how come you always write to us while driving your jeep on a rugged wagon-train trail over the Kansas prairie?"

"Jambo Cowabunga,

Thanks for your letter. I have finally deciphered it!! You ought to have become a Doctor me thinks."

And a telefax in response to a letter I hand PRINTED:

"<u>Re: Your (PRINTED) letter of 10 September:</u>

It is the first opportunity I have had of fully understanding an entire letter of yours, a rare occasion which I will cherish all my life."

ᵷ ᵷ ᵷ

My longest-running correspondence may be with Garth Thompson. His letters date back more than three decades. They are always poetic and make me homesick for Africa. For years they were in his own hand, frequently written while he was on safari, and often opened with something like, "Dear Cowabunga: As I write this, I am sitting under your favorite tree on the banks of the Zambezi, watching a bull elephant as he crosses the channel with hippos and crocodiles giving him right-of-way."

Today Garth has gone high-tech and sends me an email from his office. His words are just as idyllic, the news is just as interesting, but the printouts in my files lack the charm and adventure found in his personalized letters.

Knowing that I will respond to him with my usual handwritten snail mail, Garth closes his emails with a phrase such as, "I look forward to your letter and will get the code readers in here to decipher it."

About three months after each safari, we would have a reunion to share our scrapbooks, slides and videos. I would bring along official Cowabunga Safaris letterhead so group

members could pen a message to our respective guides. After Garth Thompson received one of these packets, he wrote the following:

Dear Gary,

How very nice to get a Cowabunga letter head with handwriting that I can read. To this I refer the delightful and kind words of all at your 1991 safari reunion.

As for your letters, well even if I can't read them I know the thought is there!! In the future just send me a letter head and have your youngest grandchild scribble around on it with your pen for a few hours; that way you can get all your letters done and we wouldn't know any different!!

Cheers,

Garth

To amuse Garth, I sat my two-year-old granddaughter at a table with a piece of Cowabunga Safaris letterhead and a pen, and told her to make whatever marks she wanted to. I then snail mailed it to Garth, just after the first of the year, thinking that would squelch him. HA! I should have known better.

A month or so later I received a snail mail reply from Garth in his customary good cheer as follows:

Dear Gary,

Please thank your youngest grandchild for the first readable letter I have ever received from you. It sounds like you really had a great time over Xmas and New Year. I'm glad she mentions how you are looking forward to your forthcoming trip to Kenya. I am also glad to see she doesn't use these irritating Pink Cowabunga papers that litter my desk, carpet, chair, pants, shirt, and brain every time I open a letter from you!!

I suggest you get your grandchild to give you a few lessons. What a pleasure—keep it up.

Kind regards,

Garth

You can't get ahead of Garth.

Every now and then I would meet someone in Africa who I felt was sophisticated enough *not* to chide me about my handwriting. One such lady was charming and intelligent, with a good sense of humor and a lovely Australian accent, *and* she was a hot air balloon pilot—the Captain of my flight over the Maasai Mara in Kenya.

On my return to Main Camp I wrote to her, sending promised photographs and a coveted Cowabunga Safaris patch she had requested for her flight suit. In time I was thrilled to receive an official Kenya Aerogramme featuring a Kenya Airways sticker that said BY AIR MAIL in English *and* KiSwahili *(Kwa ndege)*. Inside was her handwritten letter, and with each line my psyche inflated like a hot air balloon . . . until a piercing remark about my handwriting completely deflated my ego. She said:

Hello Mzee Shetani!

Fabulous to get your letter and pictures (plus Cowabunga Safaris patch—what a lucky gal I am). Glad to hear you made it home safe and that your book launch was a success (was there ever any doubt!!). Looking forward to catching up with you when next you come out to the Mara—promise me next letter you send will be typed—it's taken me 2 days to decipher your ostrich scratch!!! (Don't give up your day job as an awesome safari director!!)

Cheers, tuskers and soft landings!

Elly

I am kidding about her comment hurting my feelings—it gave me a good laugh, and she is a good friend.

Even my stateside safari friends razz me about my handwriting—sometimes in a roundabout way. One, who is a fellow author and known for his subtle humor, said: "I received your post today and ALWAYS enjoy the personal touch. I never complain about your handwriting; if I can't read it, I just take it to my pharmacy and they can read it to me."

As I draw this chapter to a close, I look at the scattered pages

of this draft, handwritten on lined notebook paper, and—ALAS! I can scarcely read my own writing, particularly when what I want to say and how I want to say it has formed in my mind faster than my fingers can write it.

Can you imagine, gentle reader, trying to decipher the entire handwritten manuscript of this book and punching it into a computerized typescript? Obviously, such a challenge would require an amanuensis of the first order. It is time to shine the spotlight on a true unsung hero of the finished product you hold in your hands: my granddaughter, Becca Wells. She not only accomplished a near-impossible task, but did so willingly while juggling college classes, family activities, and keeping up with life in general. You may have noticed the initials C.I.H.E. after her name in the book credits. Certainly she has earned the official status of **C**ertified **I**llegible **H**andwriting **E**xpert.

Thank you, Becca.

卍 卍 卍

The Bush Lingers

The Lion Sleeps Tonight— The Cheetah Hunts by Moonlight

Wait a minute! Isn't that just the opposite of what we've always been taught?

The books say that cheetahs are diurnal, seeking their prey in daylight hours on the open savannah. Likewise, despite the song, conventional wisdom has always been that lions hunt at night and sleep during the day. Turn on the TV set just now and you'll see.

So, what about those lions and cheetahs in the northern Botswana wilderness that don't seem to follow the rules?

Those who have been on safari with me over the past few decades no doubt tire of my oft-repeated phrase: "There are hundreds of generalities about animal behavior and thousands of exceptions." Never was this more true than on a Botswana safari in June 2005.

Keep in mind that winter in Botswana can be very cold, with temperatures dipping to near-freezing levels. Safarists bundled in layers of thermal underwear, fleece and wool, stocking caps, neck warmers and gloves look more like they are on an Arctic Expedition than an African Safari. Even so, when riding in an open vehicle in the pre-dawn hours, the cold wind can pierce your inner core like an ice pick.

It is with great reluctance that one gets out of bed in the "dead hour"—that time before first light in the African bush when it is darkest and coldest. Sleeping in a tent is so-o-o delicious! The air is sharp and cold on your face, while your body is warm and cozy under a heavy comforter, your feet snug against a cloth-wrapped hot water bottle. How wonderful to lie in the darkness, secure in this cocoon, listening to the sounds of the bush.

But the cycles of Africa continue in an ancient cadence, and if you don't get up then you miss out on the rhythms and the heartbeat of the day. And one day on safari—with the intensity, the excitement, the incredible array of sights, sounds,

impressions, emotions—is like two weeks anywhere else in the world. This was especially true with one particular group, the Jim Bryan family of Topeka, KS. Only four, avid, die-hard safarists, with a single focus—PHOTOGRAPHY! (No pun intended!) And they were committed and prepared to do whatever it took to achieve their goals.

Without hesitation, they would sacrifice accepted dimensions of a safari, such as campfires, sundowners, showers, sleep—even meals. And they readily tolerated bitter cold, hot sun, wind, sand, discomfort and darkness to maintain a non-stop pace to be in the right place at the right time for the best possible photographs—up to 17½ hours a day. They nearly wore out the guides and trackers but I loved every minute of it as we were in the bush for extended periods of time to experience wildlife behaviors and interactions that were new even to me!

A composite day would be up at 5:30 a.m., some quick coffee/tea and hot porridge, and into the bush before sunrise. Normally we'd return to camp at 10:30 a.m. for breakfast and then some rest. But, so much is happening we radio camp to advise "no breakfast." Finally back to camp at 2:00 p.m. for proper toilet, resupply photo equipment, etc. Lunch is scheduled for 3:30 p.m., but at 3:00 p.m. a radio call comes in—cheetahs are on the hunt! So we skip lunch, jump in the vehicle and are off, tracking cheetahs until well after sunset. Night drives with a spotlight are routine, returning to camp at 8:00 p.m. or so for sundowners and dinner. But, so much is happening we radio camp and advise "no dinner" (thank heavens for granola bars and biltong). Finally to camp at about 11:00 p.m. All is dark and silent. Quietly we creep to our tents and drop into bed exhausted, yet exhilarated.

A demanding regime? YES.

The type of safari for everyone? NO.

Nonetheless, it did result in some remarkable sightings, fantastic behaviors, and unexpected photo ops. *Example*: at night, in tall grass at the edge of a swamp, a clan of hyenas feeding on a Cape buffalo carcass reflected in the water, while crocodile eyes glow in the darkness, accompanied by a roaring chorus of frogs. *Example*: changing a flat tire at dusk with a hyena circling us. *Example*: the red dust of sunset settling in the

west while a brilliant full moon rises in the east.

So many highlights but, ah: the cheetahs. We had the privilege of spending many hours in the lives of an adult female and her two half-grown cubs. We kept a respectable distance so they accepted the presence of our vehicle while they showed us the art of being a predator. At midday they rested, then the cubs followed mother as she scouted prey. Apparently they were hungry and much of their time was spent stalking impala on the savannah—with no success. They continued to hunt after sunset and it seemed strange to see cheetahs on the prowl at night surrounded by darkness.

Then—in the spotlight behind the cheetahs—hyenas! One . . . two . . . three . . . or more! What was going on?

The hyenas were harassing the mother cheetah and trying to kill her cubs. As she was being chased she suddenly stopped, turned, and confronted the lead hyena. Then she jumped in the air and swatted the hyena in the face with her right front paw! WOW! Catching up with her two cubs, all three faced the hyenas and hissed while in a defensive posture. It was a stand-off, but the cheetahs survived this particular encounter with their enemy to hunt another day.

In contrast to this drama was the night we came across a pride of ten lions not long after dusk. They were asleep on a mound which put them at eye level with us from our elevated position in the open vehicle. The lions were unperturbed by our spotlight which resulted in a great photo op of various sleeping postures. We decided to stay with them in hopes they would hunt later. Our spotlight was powered by the vehicle battery, and we had to use it sparingly. It was dark.

We waited.

By moonlight we could see when the lions shifted positions, and we would then check them briefly with the spotlight.

We waited.

It was cold. The lions were huddled together in a pile of warmth.

We waited.

An older female awakened, groomed a bit, and went back to sleep.

We waited.

It was quiet. We could hear each other's stomachs growling . . . .

We waited.

The lions slept. Minutes passed.

We waited.

The lions slept. Hours passed. No one talked. You might think this would be boring, but there was something magical about being in the bush at night, in an open vehicle, under the Southern Cross and intermittent shooting stars, *with a pride of lions*— just ten meters away in the silvery moonlight.

We waited.

They slept. Then we heard an unfamiliar noise. I can usually identify the sounds of the bush, but what was this? We searched around us with the spotlight . . . nothing. It was coming from the direction of the mound—listen . . . Ohmigosh!

The sound was a *lion snoring*!

They slept.

We waited.

The time must have been approaching midnight and our vehicle battery was getting low from spotlight use. We finally gave up and returned to camp—but what a privilege it had been for all of us.

I was most impressed with the dedication and passion of the safarists. Within this unique team of four photographers, one of them took general scenes and kept detailed notes, another shot over 25 hours of sound video, and the two with super-sophisticated digital cameras took innumerable still photos—in fact, nearly 60,000 images between them! As for me, I didn't take a camera, but reveled in absorbing as much of the African wilderness into my spirit as possible.

Weeks after my return to the USA I was still finding Kalahari sand in my shoes, my hat, my field bag, my dopp kit, my binoculars, and even my journal. But I didn't mind.

It meant that the African bush was still with me, everywhere.

᭡ ᭡ ᭡

Isiolo, Kazangula, Bujumbura & Beyond

Evocative names that dance on your tongue:

Isiolo (ee-**see**-oh-low)

Kazangula (ka-**zahn**-goo-lah)

Bujumbura (boo-**juhm**-burr-ah)

The romance of Africa—its mystery, excitement, and remoteness from everyday life—is epitomized by such legendary destinations as Isiolo, Kazangula and Bujumbura, as well as Jinja . . . Bulawayo . . . Upington . . . Iringa . . . Keetmanshoop . . . Bagamoyo.

These exotic locales, along with Ujiji, Zanzibar and Timbuktu, are emblematic of the Africa of yesteryear, and the essence of each is indelibly imprinted upon the passport of my soul.

卍 卍 卍

Isiolo

Just the word itself is alluring, enticing.

Isiolo. I would pinpoint its location on my Kenya maps.

Isiolo. I would read about it in the literature of Old Africa.

Isiolo. I would dream of going there someday, and my longing would intensify.

How fortuitous that destiny took me to Isiolo on my first safari in 1974. From Nairobi we headed north, on the main road from Cape Town to Cairo, crossing the equator and climbing through the agricultural highlands. I followed our progress on a map, noting landmarks in my journal and periodically recording our altitude, which culminated at over 7,000 feet.

My altimeter showed we had begun a gradual decent when abruptly, the terrain dropped to flat, open country extending north to distant, rugged mountains. At that moment I had my first view of Isiolo—dramatic, unexpected.

Goose bumps!

It seemed to take forever to get to the edge of town, and

en route the earth changed from rich, colorful soil to semi-arid scrub and dust. The moist, cool highlands air became hot and dry, and the sun glared intensely. Finally, we reached the outskirts of town and slowly crawled past a series of small buildings painted blue or green or white, including bars, shops, stores, bicycle repair garages, and . . . the Little Angels Nursery School (such a lovely name).

Proceeding through town, we passed two prominent structures—an ornate Muslim mosque and a Catholic church under construction—and a whirlwind of humanity: children playing, women socializing, men bartering—some people in traditional dress, others in well-worn western clothing. At the market in the center of town, we saw many of the ethnic groups from the desert region of Kenya: Samburu, Borana, Gabbra, Rendille, Somali, Turkana—a veritable crossroads of culture.

And, of course, the omnipresent animals: dogs chasing their tails and each other; goats foraging on vegetation stubble and rubbish; burros carrying huge sacks of grain and heavy jerrycans of water; camels hobbled and bellowing; cows lying in the middle of the road, nonchalantly chewing their cud; small birds flitting around in search of crumbs or seeds.

At the north boundary of town, we made the required stop at the checkpoint for the NFD—the Northern Frontier District, a wild and lawless area inhabited by *shifta* or marauders. It was a simple wooden shed manned by armed military personnel, and each driver had to register such details as vehicle license number, nationalities of passengers, destinations, and scheduled date of return.

Approaching the checkpoint, our driver had wisely advised us to shut our windows. Even before we stopped, a horde of over-aggressive vendors surrounded our vehicle, trying to sell everything from jewelry to carvings, to bananas, to daggers with wicked blades. Four to five deep, they jostled for a better position, knocking on our windows while dangling their wares at eye level. All of this was a bit much for some in the group, but I found it to be a fascinating experience. Our driver finished with the formalities and we were off on our own.

On our return several days later, we were again at the

checkpoint. This time, with permission from our driver, I exited the vehicle and sat down beside it. Immediately the swarm of vendors crowded around me like vultures on a kill, each offering "a good price." I raised a hand, spoke what little KiSwahili I knew, and introduced myself. Apparently no one had done this, as they all were quiet and curious. Then I asked about them, admired their offerings, bought some bananas from a woman with a baby on her back, said goodbye, got into the vehicle and closed the door. HA! Instantly they converged around the vehicle in a herd mentality, showing their initial behavior. On successive journeys through Isiolo, I became better acquainted with many of the vendors.

One of the more anomalous adventures to befall a group had closure in the NFD. It was always my practice when traveling to Africa *not* to check luggage, and take only a carry-on bag. On one trip to Kenya, for whatever reason, I checked my bag along with everyone else. On arrival in Nairobi we all marveled as a plethora of baggage coming off the 747 went round and round on the carousel. But my bag never showed—nor did the bags for anyone else in my group! The *entire* lot of luggage for the fifteen Cowabunga Safarists was lost—a first (and thankfully the last).

The airline officials had absolutely no idea where our bags were, but assured us they would arrive on tomorrow's flight. Meanwhile, "Please fill out this *Lost Baggage Report*, together with your routing, and we'll deliver your bags on to you." It had been a long series of flights from the USA, it was late, everyone was tired and jet-lagged, and few could answer all of the questions on the form about the style, shape, make, color of bag, etc. Finally, we were off to the hotel with just the clothes on our back and our carry-on bags, mostly camera equipment.

The next morning, the group was eager to hit the safari trail and resigned to doing without their bags for a day or so. And was I ever glad that *my* bag was lost with all the others.

Our first day on safari was marvelous, but the bags did not show up as promised. Nor did they the next day . . . or the next . . . or . . . . Initially this was of concern, but the group quickly adapted and observed how carefree they felt not to be burdened with all those material possessions!

Tee shirts washed at night were dry by morning. So that everyone could have a different look each day, we would switch shirts with one another, and hats, and sunglasses, and whatever. It made for a hilarious fashion show that always required a new round of photos.

Our camp in the NFD was hot and dusty, but the spirit of the group was strong. We were well into the safari mode and our sightings had been superb, but the missing bags preyed on my mind. I felt bad for the group, but no one complained. After seven days with no word, I felt helpless and frustrated.

The next day after lunch, everyone retired to the shade of their tent for a rest. I was in front of my tent, concentrating on journal entries and my wildlife checklist. One of the camp staff suddenly appeared at my side, pointed to the south and said: "Cowabunga, look! It's the luggage!"

I gazed across the flat, dry scrubland, through the shimmering heat waves to the far horizon, and noted a funnel of dust trailing to the sky. It was not whirling and dancing like a dust devil, but making a beeline in our direction, as though following a vehicle. I focused in with my binoculars and—YES!—it was an undersized, overloaded truck, inching its way toward camp.

What a great relief this would be for the group! When the truck arrived I coached the camp staff to unload quietly, and we neatly placed the luggage by the campfire circle. Then excitedly I ran up and down the tent row, announcing, "Cowabunga! The bags are HERE!"

The group emerged from their tents, shaded their eyes, looked at the bags, shrugged their shoulders and said, "What do we need these for?"

ᕤ ᕤ ᕤ

Kazangula

Kazangula. Is there such a place?

Kazangula. It sounds fictional, like something from an old Tarzan movie.

Kazangula. It is *not* make-believe—it is real and verifiable, a vibrant, pulsating border post on the banks of the Zambezi in Botswana, across from Zambia.

At Kazangula there is constant scurrying, noise, disorder and confusion, but somehow it works. People cross the river from one country to the other in everything from primitive dugout canoes to fancy motor boats. Autos, buses, and heavy transit-goods trucks—some with triple trailers and as many as 28 wheels—cross on overcrowded ferries . . . *when* they're running.

And Kazangula lays claim to a one-of-a-kind geographical feature: the world's only quadrapoint. It is the singular point on the planet where the borders of four nations touch: Zimbabwe, Zambia, Botswana, and Namibia (because of the Caprivi Strip).

If you did not know about this extraordinary idiosyncrasy

before you arrived at Kazangula, then you might never know about it—even *after* you arrived. There is *no* marker, *no* plaque, *no* flag, *no* photo-point—or *any* other indication of the location of the world's only quadrapoint.

Why not?

Inasmuch as the four corners of the countries' borders meet *underwater*, in the middle of the Zambezi, there is no obvious sign.

Whenever I had a group crossing the river at Kazangula, I would brief them about the quadrapoint, advise them to have their cameras at-the-ready, and instruct the boatman to make a slow circle in the middle of the Zambezi. Then I'd point to the exact spot and tell them to, "Get that shot."

I dare say that only Cowabunga Safaris groups have such a photo, as well as a certificate of documentation *(see sample on next page)*.

卍 卍 卍

COWABUNGA
S A F A R I S
Africa under a rainbow

HEAR YE ! HEAR YE ! HEAR YE !

Be it known to all living creatures (and the spirits of those great African explorers who have set foot upon the continent) that your humble and abiding servant

_____

has successfully navigated and traversed the world's only QUADRAPOINT where the borders of four nations – Namibia, Zambia, Zimbabwe and Botswana – touch in the middle of the mighty Zambezi River.

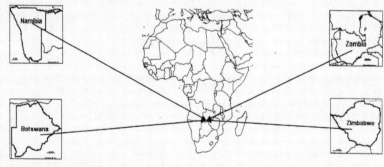

Attested on this date _____ *at Kazangula by my signature.*

_____
Gary K. Clarke,
President for Life, Cowabunga Safaris

Bujumbura

Bujumbura. The capital of the tiny country of Burundi, just south of Rwanda.

Bujumbura! I was on a direct flight to Entebbe, Uganda when the Captain announced we would soon be landing in Bujumbura.

Bujumbura? I had never been there. I had never thought of going there. And now, an unscheduled stop with no explanation. How exciting!

Often, in this type of situation, passengers are confined to the plane. I asked if I could disembark and was granted permission and given an IN TRANSIT card to re-board.

The airport was nondescript, with a strong presence of uniformed, armed security. A small curio counter was devoid of anything reflecting a Burundi origin. I did, however, find three small totem-like figures depicting the "see no evil, hear no evil, speak no evil" catchphrase, carved from bone.

With this little treasure in hand, I moseyed to the ubiquitous bar and had a local beer—warm, of course. Then an airport

official advised me to re-board. I was delighted to have set foot in Bujumbura, a somewhat obscure pinpoint on the vast African continent. I hoped at the time that my next serendipitous adventure might take me to the esoteric village of Gondokoro, on the banks of the White Nile . . . and I'm still hoping.

〓 〓 〓

And Beyond: Equator to the Cape

In September and October 2003, I had a safari extraordinaire. From my journal I will share some highlights, written here as they originally were, in present tense.

I began this adventure just north of the Equator, with the glaciers of Mt. Kenya ever-present. At its end I write this in Cape Town, while misty clouds roll off the edge of majestic Table Mountain.

After five weeks in the bush on back-to-back safaris in Kenya, Botswana, Namibia, Zambia and South Africa, I've experienced fierce winds, blazing sun, swift rivers, blowing sand, intense heat, choking dust, and rough, ocean swells.

In the course of the journey I:

—was greeted unexpectedly at the Nairobi Airport by 15 Kikuyu in traditional dress, with a celebration of drums, songs and dance.

—floated in a hot air balloon over the Great Migration.

—participated in a surprise ceremony in the bush to reconfirm my status as a Maasai Elder in the Oltukai Mara Clan.

—donated cases of education supplies to elementary students at a remote school for Maasai children.

—explored the Shetani Lava Flow, the source of my KiSwahili name, under the shadow of Kilimanjaro.

—tracked wild dogs in the Okavango as they made two kills.

—watched the sunset over the Okavango Delta while on the water and drinking sundowners.

—marveled at a pride of 22 lions on an <u>elephant</u> kill, then lived in fear while the pride terrorized our camp for three days and nights.

—had three bull elephants swim perilously close to our boat

when crossing the Chobe River.

—unexpectedly entered Namibia at an isolated border post in the narrow Caprivi Strip.

—crossed the Zambezi River at Kazangula in a small boat.

—arm wrestled the "World's Strongest Man" in Livingstone, Zambia—and lost!

—stood in the center of the 100-year-old bridge over the Batoka Gorge between Zambia and Zimbabwe, and felt it shudder as a train crossed over the bridge past me.

—flew in a Microlight over Victoria Falls and the Zambezi River.

—followed, off-road and in an open vehicle, a dominant male leopard as he patrolled and marked his territory in rugged bushveld.

—navigated the turbulent, cold Atlantic Ocean off the coast of South Africa to reach the feeding waters of the great white shark, and experience remarkable sightings just arm's length from our boat.

—thanked the safari gods for such unexpected and fantastic eventualities.

In just a few hours, I'll board a jetliner for an overnight flight *up* the entire length of the continent—from Cape Town to Cairo and beyond.

I am severely weather-beaten, sunburned and windblown, physically exhausted and emotionally drained. But it was incredible Africa at its finest.

꒰ ꒰ ꒰

A Tusker Tribute

An old bull elephant with large tusks is known as a "tusker."

One of the beers produced by East African Breweries in Nairobi is called Tusker, and it is Kenya's largest-selling beer. It was named in memory of the founder of the Breweries, Charles Hurst, who was killed by a rogue bull elephant while on safari.

During my 25th Anniversary year of leading safaris, extensive celebrations were held throughout Africa (any excuse for feasting, singing, and dancing). I was honored and humbled, and

a good time was had by all.

Apparently East African Breweries had word of my fondness for Tusker beer—yes, I even brush my teeth with it—over a quarter-century. Hence, their tribute to my African travels was truly distinctive.

We were at the end of a rather intense and somewhat exhausting Kenya safari. My group, tired but fulfilled, had just completed the exit formalities at Jomo Kenyatta International Airport, so we all proceeded to the departure lounge. And there they were—representatives from East African Breweries!

With great panache they presented me with a variety of Tusker oriented items featuring the elephant logo, and the classic black, yellow and white colors: a bottle opener, a small banner, a cap, a ball-point pen, a bar towel, a set of coasters, etc. All were appreciated. The *piece de resistance,* however, was . . . a Tusker tire cover for the spare on my Jeep! It is fantabulous!

It sets my Jeep, the Blue Dung Beetle, apart. And, it is the *only* Tusker tire cover in Topeka . . . in Kansas . . . in the USA . . . heck, probably the only one outside of Africa!

To top it all off, the good people of East African Breweries presented me with a case—yes, a **CASE**—of Tusker beer! A lovely gesture, but I could not take it on the plane, and we were scheduled to board soon. I looked at my group, my group looked at me, and without a spoken word, we all knew that we had no choice but to drink the entire case, then and there. Cheers!

We were flying high before the plane took off, and there were numerous toasts to Isiolo, Kazangula, Bujumbura and Beyond.

ᚅ ᚅ ᚅ

Crocodiles in Trees?

Crocodiles in trees? Yes. I could not believe my eyes!

In May 2004, I made my first venture to the country of Gabon in Central Africa. Sometimes called "The Edge of Africa," Gabon has a coastline where the equatorial rain forest meets the Atlantic Ocean. The country supports a mosaic of forests, savannas, and mountains that contain lowland gorilla, forest elephant, mandrill, chimpanzee, red buffalo, and a wealth of other species. A series of *National Geographic* articles (October 2000, March 2001, and August 2001) followed Michael Fay's Megatransect in Gabon, across 2,000 miles of untamed Africa, all on foot. Fay then convinced Gabon's President, Omar Bongo, to establish 13 national parks for conservation and tourism.

Researching my antiquarian map collection, I came across a classic map by the famous explorer Richard F. Burton of what is now Gabon.

Published in 1876, it is simply titled *THE LANDS OF THE*

GORILLA. Mountain ranges, bays, rivers, even villages are shown, but most of the designations are simply GORILLA COUNTRY. One, however, does indicate a tribal group and says: "MOSHOBO? (Said to be cannibals)."

During one treacherous trek in Gabon, we went in search of lowland gorillas. We were alert and cautious—not because of gorillas, but because of the possibility of meeting elephant or buffalo in such close quarters as their spoor was everywhere. The path, when there was one, was undulating and uneven.

After many hours, we heard and smelled gorillas and chimps, but did not see them. Yet we did get sightings of several monkey species I had never seen in the wild: mustached guenon, putty-nosed monkey, grey-cheeked mangabey and red-capped monkey.

Later we were permitted to trek on "Gorilla Island," where a family of them had been introduced. Here I was able to observe, as well as photograph, gorillas. Also on the island were captivating orphan gorillas being hand-raised by researchers.

Gabon offers unique sightings and unusual behaviors. While I did not see elephant on the beach or hippo swimming in the

Atlantic ocean (as reported in *National Geographic*), I did see spot-necked otter, sitatunga, and manatee. And our excursion up the Mpivi River was fantastic. The river, a narrow waterway with a wall of trees forming a gallery forest towering to the heavens on either side, is a "Heart of Darkness" setting. The water is **BLACK** and the brilliant green vegetation creates mirror-image reflections. I saw more African finfoots, a seldom-seen water bird, in one afternoon than I had on all my previous safaris.

But the most surprising oddity was—believe it or not— crocodiles in trees! Towering trees had grown at an angle toward the center of the river, probably seeking light. From a distance there would appear to be a H-U-G-E monitor lizard resting on a tree trunk high above the water. No, wait—through binoculars it looked like a slender-snouted crocodile—in a tree! As we got closer, it was! And then—SPLASH!

An adult crocodile would scoot off the tree and belly-flop into the river just ahead of us. And then another! I had never seen this behavior previously, or even a reference to it in the literature. Why do *these* crocs manifest this peculiar quirk? I have developed a theory, which would be an ideal topic for sundowners sometime. You are invited.

My adventures in Gabon were a bit perilous at times, but a new dimension of Africa. The most dangerous part of my journey was in Paris, where I made a connecting flight from the USA to Libreville, the capital of Gabon. Due to airline schedules, I had a twelve-hour wait at the Charles de Gaulle Airport. I spent most of it comfortably seated in the spacious, new, ultra-modern, multi-million-dollar terminal I had read about—the architectural pride of France.

Now and then, I would pause in my reading to glance up at the patterned, high-vaulted ceiling. I noticed that outside light was conspicuous through tiny cracks and wondered if the ceiling leaked when it rained (surely not), or even if the entire roof was structurally sound (surely it was).

I continued reading, but couldn't help looking up every once in a while . . . and wondering.

Four days later while I was in Gabon the roof collapsed and killed five people.

Sunrise . . . Sunset

A *blazing* sunrise . . . a *fiery* sunset.

Nature's longest running spectacle occurs over much of the earth. Only the North and South Poles are immune to this daily natural phenomenon.

Granted, a sunrise or a sunset may not always be apparent. The sky may be totally overcast. Or the sun might be blotted by clouds, smoke, or dust. Still, there is a sense of when the day begins and when the day ends.

What is it about a visible sunrise—or sunset—that is so enchanting?

In my opinion, there are several facets. Most obvious is the natural beauty. These solar showcases offer an unparalleled visual display of colors and hues.

Another is the cycle of life. Living beings—plants, animals, humans—respond to the cycles of day and night, light and dark, warm and cool. Sunrise and sunset keeps these systems in sync with the rhythm and heartbeat of the earth.

Sunrise . . . Sunset.

Each word seems so simple, yet implies complexity.

Sunrise . . . Sunset.

Are they symbolic of new life, or temporary death? Of a new beginning, a fitting end? Of an opportunity to renew, or to reconsider a failed attempt?

I am not sure.

Certainly sunrise signals the beginning of a new day, and sunset the completion of the diurnal cycle. But there must be something more . . . .

Perchance, is there meaning beyond our conscious level? A candle burns brightly, then fades to a dim glow. A gazelle at birth has life and vitality, and later becomes old and decrepit. A tree starts as a perfect sapling, but eventually is leafless and shattered. Could it be that in our subconscious, these beginnings and endings are reflected in sunrise and sunset?

In Kansas, the late-evening summer sunsets are long and

lingering, but in mid-winter sunrise and sunset can occur so quickly that I may not even notice. When this happens, I am always disappointed and often provoked with myself. To be so preoccupied with the "business of the day" that I miss a sunrise or sunset is inexcusable! These are two of the great joys of daily life—available and free.

Africa offers a different scenario. I am more in control of my time. I am living life as if should be. No need for an alarm clock. With the morning, I sense when to awake just at first light, especially if I'm sleeping in a tent . . . or better yet, in the open under the stars.

In Africa, bisected by the equator, the alpha and omega of each day has particular significance, with consistency in length and timing, and little variance from season to season. For me, the most spectacular African *sunrises* are those in an expansive, cloud-streaked sky, with a radiant burst of gold over the dark pinnacle of a volcanic mountain. The most spectacular African *sunsets* are those when pronounced dust is suspended in the air and catches scattered rays of light to create a red blaze, like a bush fire on the horizon.

Nature fashions a brilliant jewel every morning, and paints a new masterpiece every evening.

With the evening light of Africa, which is magical, I automatically pause at sunset. To me, it is not simply an external event, but a happening deep within myself, a dimension of awareness beyond my human perceptions. It is like a mystical communion with the bush and all its creatures.

Sunset on safari is such a compelling moment that I cannot let it pass. Regardless of my circumstance, I simply must acknowledge it, preferably with a sundowner. The drama and splendor of the setting sun often elicits conservation among my safari companions. But just before the sun touches the horizon all talk ceases, and in silence we watch the sun vanish.

Remarkable.

Writer Keith Bellows summarized it this way: "Sundowners are a tradition that recognizes the rhythm of the land, the cadence set by the sunrises and sunsets that govern the life of the wild. And that, in Africa, governs us, too."

Sunrise . . . Sunset

If you've been on safari, then you know. If not, make this dawn and dusk phenomenon an intrinsic part of your daily living.

卍 卍 卍

The Lady in Line

"Aren't you Gary Clarke?" asked the lady behind me in line at the post office.

"Yes, I am."

"My," she said with disbelief, "I didn't know that you were still alive!"

Not knowing whether to laugh or cry, I just smiled.

"I remember when you were at the Topeka Zoo, and you were on TV all the time," she continued. "I've been to many of your safari lectures and seen your wildlife photos from Africa. My, you have had an adventuresome life."

She paused, and then asked, "How old are you now?"

"I'm 74."

Glancing at my cane and carefree attire, she said, "Oh; I guess you are just coasting in retirement."

After we both had completed our transactions, we met in the

lobby and I took a moment to tell her my current status.

"Actually, I'm at the pinnacle of my third lifetime," I explained. "My first lifetime was zoos; my second lifetime was safaris; and now my third lifetime is writing books and keeping up with ten grandchildren. I am halfway through the draft manuscript of my fourth book right now."

"Oh my, I guess you'll never really retire," she said. Then she wished me luck and was on her way.

This chance encounter did give me pause to consider how fascinating life's journey has been for me. Please permit me to share a few reflections with you.

When I was a boy, there were two things I wanted to do: work in a zoo and travel to Africa. I was living in Kansas City, Missouri, when I started my zoo career as a keeper at the Kansas City Zoo at age 18. Later, I conducted research on reptile biology at Midwest Research Institute in Kansas City, where I experienced a near-fatal bite by a red diamond rattlesnake, which was featured in *The New Yorker Magazine,* June 21, 1961. At age 23, I became a curator at the zoo in Fort Worth, Texas, and then one year later director of the

Topeka Zoo in 1963.

Thanks to tremendous community support, what once was the Gage Park Zoo evolved into the World Famous Topeka Zoo. You may have read about Operation Noah's Ark, the first giraffe born in Kansas (Sunflower), or the opening of Gorilla Encounter. Djakarta Jim, our orangutan artist, was the cover story of the October 1971 issue of *Science Digest,* and our gorillas, Max and Tiffany, were featured in the *The New Yorker Magazine*, August 16, 1982. Our zoo received a conservation award for the first successful hatching of an American Golden Eagle, and an exhibit award for the Tropical Rain Forest. In 1971 I was elected the first president of the independent Association of Zoos & Aquariums. In 1975, President Jimmy Carter appointed me to the newly formed National Museum Services Board to represent *all* zoos and aquariums in the USA. Our zoo celebrated its 50th Anniversary in 1983 with a visiting white tiger, and qualified to borrow a koala from the San Diego Zoo. Whenever San Diego called about making the arrangements, they would say, "This is the *other* world famous zoo calling."

Zoo life was rewarding and fulfilling. My dream of going to

Africa also came true with a safari in 1974. How lucky could I be? To work in a zoo and go to Africa, too!

In 1989, I achieved a long-time goal and climbed Kilimanjaro, the highest mountain in Africa. I turned 50 on the mountain, and decided to make a dramatic career change: after 26 years as Topeka Zoo director, I gave up that career to devote my time to sharing Africa with other people through my own company—Cowabunga Safaris.

By 2006, I had completed safari number one-hundred-and-forty. A degenerative nerve condition in my legs precluded me from actively leading more safaris. This, in turn, led to an unexpected transition into my third lifetime—writing books.

I had always enjoyed *reading* books . . . and I had had some thoughts about *writing* books as well, but I knew that would be time-consuming.

As a zoo director, my life had been 24/7 **ZOO**. As a safari leader, my life was non-stop **AFRICA**. In both instances, it was my choice, and I relished each calling with great enthusiasm and profound gratitude.

But to do *serious* writing would require that I stay in one place and take the time necessary for creative reflection and wordsmithing—time to cogitate, to intellectualize, to muse, to formulate expression. It appeared that time was NOW. My limited mobility presented a golden opportunity for me to get words from my mind through my fingers and onto paper.

Do I miss Africa?

Indeed.

Yet Kansas, my birthplace, has some fundamental characteristics of Africa: an all-encompassing sky . . . dramatic clouds . . . distant horizons. From my home in Topeka, it is a short Jeep drive to the Flint Hills, where I am able:

—to relish the tranquil solitude.

—to bask in the same sun that warmed me on safari.

—to gaze upon the same moon that I knew in Africa.

—to hear the lyrical whisper of the wind.

—to ponder rain . . . and fire . . . and space . . .

On a much smaller scale, my Topeka backyard serves as a

small window of nature in the heart of town. From my deck overlooking the backyard I am able to:

—behold stately trees dappled with sunlight or blanketed in snow.

—observe the expressive colors of the changing seasons.

—watch fascinating native birds and listen to their song.

—follow the lives of small animals and their families.

—track theatrical thunderstorms.

—thrill at the fiery spectacle of a Kansas sunset.

—be enchanted by dancing snowflakes.

If, perchance, the lady in line reads this, then thank you. I wish, when we visited at the post office, that I had thought to share with you one of my favorite quotes:

If time were measured by the events that fill it,

I have lived two hundred years.

—Beaumarchais

Give Up Africa? Me?
<u>NEVER</u>!

So here I was, my crutches in my lap, being pushed in a wheelchair through Washington-Dulles International Airport to rendezvous with my safari group—most of whom had never met me. Shock registered on their faces at my approach. One man elbowed his friend and I overheard him say: "You mean THAT is the guy who is going to lead us through Africa?"

It was then I realized that soon I had to make the long-delayed "decision."

Most of my current safaris had been booked by old friends who knew about my degenerating physical condition and limitations. "We just want you along for your bad puns and corny jokes," they would say.

Maybe so, but I felt it was only fair for future groups to have an active safari leader capable of looking after *all* of the

details, sorting out the inevitable problems along the way, and orchestrating spur-of-the-moment experiences to ensure that the safari would be the most memorable travel experience of their lives. As the safari leader, I felt it was important to be the first one up in the morning and the last to bed at night. For decades I did just that with ease; now it was a struggle. In addition to my general health issues, the countless long flights over the years were taking their toll and draining my energy.

And there was something else: when I lagged behind the group on safari, they would express concern and offer to carry my camera bag, or help me up the steps, or ask if I was OK. While I was appreciative, I felt uncomfortable because the focus of attention was on me rather than on Africa.

Hence, the "decision" facing me straight away was whether or not to *personally* carry on leading each safari.

In my mind, I knew that the right decision was to forgo that great privilege. But in my heart . . . could I give up Africa? *Me?* **NEVER!**

Africa, in mind and in spirit, is part of my very existence—

with each thought, each breath, each heartbeat. And there is such joy in sharing.

A safari is a travel experience unlike any other—bringing together a group of people who may or may not be previously acquainted, but eagerly anticipate experiencing a unique adventure together.

Instantly, there is a common bond—Africa and its wildlife—that brings a bright and happy spiritual substance to the group. People from various backgrounds feel similar qualities, and drop the facades of society to be on a comparable and compatible level with other group members—many of whom are strangers, but often become lifelong friends. That's what makes a safari so special.

Having first-timers on safari was always a delight. They were *so* excited—adults agog with childlike anticipation. They had prepared for months and knew more than most others about Africa. Still, it was satisfying for me to see Africa herself dispel, in a predictable pattern of reactions, the persistent misconceptions so many people have about the continent.

Most African airports lack an enclosed jetway, so passengers exit the plane into the open air to descend mobile steps. This elicits their first comment: "My, this weather is wonderful." I guess they expected the hot, steamy jungles of Hollywood movies.

Inside the terminal, they met my African friends, safari drivers and guides. Their second comment: "Oh, these people are so gracious." I guess they expected a menacing warrior with a spear.

Then on to camp and our first meal, with the comment: "Why, this food is delicious . . . and healthy." I guess they expected dehydrated rations.

After our initial game drive, they remark: "WOW! I didn't think we'd see this many animals on the entire safari, let alone the first day."

By then, there was the unanimous exclamation: "If I had known Africa was like this, I would have been here years ago!"

I remember one man who demonstrated the classic mindset of a safari novice. We were enroute to our first camp and still quite a distance from the national park. Suddenly he shouted, "Stop-Stop-**Stop-Stop-STOP-STOP**!" The driver slammed on the brakes, and I asked, "What's the matter?" I thought he was ill, or his camera had bounced out the window, or . . . .

"Look!" he said, pointing in the distance. On the far horizon was a zebra. "It's my first animal on my first safari," he said. "I have to get a photo."

"Why don't you wait?" I suggested. "We'll be much closer with better photo ops soon."

"NO!" he replied. "I must get this picture NOW." Even with his telephoto lens, the zebra was so far away that it would be a mere speck on the horizon. But I understood how he felt, and knew it was important.

In the course of the safari, we *did* have great photo ops of zebra and other species. At one point, we were in the midst of the great migration, driving slowly through mass herds of wildebeest and zebra extending as far as the eye could see. They parted like the Red Sea around our vehicle, so close you could reach out and touch them. "Now," I said, "*this* is a great

zebra photo op." In a low, dispassionate voice, he replied, "I've already got so many pictures of zebras I wouldn't know what to do with any more."

I loved it when—just before our dawn game drive—the first-timers would ask, "What are we going to see?" Expecting me to give them a laundry list of animals, they may have found my answer disappointing.

"I'm not sure," I would say. Then I'd explain.

"I can tell you what species are found in this habitat, what animals Cowabunga groups have seen on previous safaris, sightings reported by others from the past few days, the location of waterholes, dens and nesting areas, as well as the influence of weather. But—if I *knew* what we were going to see on this game drive, I would not be here. If I want edited, spoon-fed nature, I'll watch a wildlife video. If I want repetitive, fabricated nature, I'll go to a theme park. But what I want is *true* nature—spontaneous and unpredictable. THAT is why I am *in* Africa, *on* safari, at *this* wilderness locale. So let us venture forth, together, to see what the bush brings us."

I know of intensely managed private reserves in Africa where you can easily see the "Big Five" before breakfast. To me, that is *not* a safari.

The singular animal—so symbolic of Africa—that everyone on safari feels is a "must see" is the lion. Yet I've been on many exhilarating and memorable safaris with no lion sightings, and the groups were so fulfilled anyway they did not realize it until I reminded them.

There are some people I met for the first time on their one and only safari, and I've never seen them again—yet more than 35 years later they continue to send me news clips on Africa, articles on wildlife, books and magazines, and holiday cards with notes expressing how meaningful their safari was to them.

So gratifying to me are the many letters from safarists. With their permission I'll share one from the Sutcliffe family—Joe, Rita and their young son Nicholas, who earned the nickname CJ for Cowabunga Junior.

June 25, 1996

Gary,

Knowing you would be on your way to Tanzania about the time we
arrived home from Amsterdam gave us a little extra time to gather
some of our thoughts about the wonderful trip and memories you
provided for us.

Gary, we know our trip with you to Kenya will be the most
remembered trip of our lifetime.....unless we ever go to the
"moon"! It's not possible to adequately express our thoughts,
emotions, and exuberance about our "adventure" with you and our
unending thanks to you for making it more special than we ever
thought possible. There were many highlights during our
adventure, but when summarizing our thoughts we find so many of
the highlights would have gone unnoticed if we had travelled with
anyone else but YOU.

Your spirit of respect for all people, the African customs and
culture, and Mother Earth will be a lasting memory. Respecting
the African traditions and way of doing things....speaking their
language....not making comparisons with other countries or the
United States....knowing their way of thinking and trying to think
the way they do....being an outstanding ambassador and teacher.
These all describe our memories of you, and perhaps one of our
most memorable thoughts about you, Gary, is you are "authentic".
In Gary Clarke, "what you see is what you get"....honesty,
integrity, knowledge, humility, caring, patience, understanding,
respect and, most of all....FUN.

WE THANK YOU FOR MEETING AND EXCEEDING ALL OUR HOPES AND
EXPECTATIONS!

With Special Memories,

Joe, Rita and Nicholas (CJ) Sutcliffe

Over the years, I developed so many friendships in Africa that had I wanted, I could have been a "bush bum," living off of a steady stream of invitations from camp managers to be their guest, from guides to tag along on game drives, even from locals to stay in their home. On occasion I did so, with sincere appreciation to my gracious hosts. And while I enjoyed myself immensely, something was lacking. *I missed my groups!* Despite the momentous responsibilities, the rewards of sharing with others are for me unparalleled.

As I reminisce about these magical days on safari, the memories are vivid and everlasting: the roar of the lion, the cry of the fish eagle; sunlight and thunder, mirages and rainbows; shadows and dust, reflections and silhouettes; campfires and camaraderie, laughter and song; love and spirituality, respect and appreciation.

The journals from my 140 safaris occupy nearly eight feet of shelf space in my library. As I thumb through them, I find a lot of Africa in the pages, far beyond words: leaves, scents, sand, insects, personal notes, postage stamps, beer labels, entry/

departure forms, thorns, bank notes, and airline barf bags. Back in Kansas, these journals provide a tangible link to my life and my years in Africa, and the influence it had on who I am today.

By birth I am a son of Kansas, by choice a son of Africa, and in spirit, I belong to both.

My interactions with the indigenous peoples of Africa gave me an intensified sense of spirituality and a deeper respect for the hidden realms of other cultures, which led to another perspective of Western values and way of life.

For a kid from Kansas who revelled in the animal kingdom by means of books and zoos, who emulated Martin and Osa Johnson through their travels and photography, who dreamed of Africa via maps and museums, indeed I have been most fortunate.

Though I have not been on safari since 2006, visions of Africa are forever with me:

—the smoke and thunder of Victoria Falls.

—the green immensity of the Serengeti.

—the intrigue and romance of Timbuktu.

—the soaring, cloud-wreathed volcano that is Kilimanjaro.

—the excitement and grandeur of the Great Rift Valley.

—the savage sparkle of the Zambezi.

I *crave* Africa.

I *ache* for Africa.

If I am unable to return, I find solace that Africa is a <u>vital</u> element of my being—so much so I need not physically be on the continent for it to revitalize my spirit.

Give up Africa? *Me?* **NEVER!**

Safari is a State of Mind.

卍 卍 卍

...a faint cry...

...from deep...

...in the...

...African bush...

SEND

CHOCOLATE!

Cowabunga...

Acknowledgements

Special thanks to Becca Wells, C.I.H.E., for her time and effort deciphering my illegible handwriting from the original manuscript, as well as her clever formatting of the initial computerized text. I am indebted to Kay Quinn for making my rambling stories readable; you, the reader, are the beneficiary of her superb editorial skills. From the outset I wanted this book to be "different" from my other books in format. The artistic photographic techniques of Rod Furgason and the professional skills of Mary Napier brought my concepts to fruition. Joe and Rita Sutcliffe were always ready to help with special projects, Randy "Flatdog" Austin was steadfast in his role as editorial advisor, and Gary Lee contributed original sketches for the book. Asante sana to one and all.

My appreciation, as well, to the following for assistance and support: Brian Hesse (aka Mzungu Mrefu), Ken and Jane Yocum, Rod and Jan Furgason, Garth Thompson, James & Janine Varden, Sarah Lamb, the Cumings Family, Mzee

Schuyler Jones, CBE, Dick Houston, Jim Bertoncin, Neta Jeffus, Steve & Kathy Clarke, Ken and Margie Blanchard, Sherry Best, the Topeka & Shawnee County Public Library, Lloyd Zimmer Books & Maps, Dennis Baranski, Phil Grecian, the Jim Bryan Family, Burt Propp, Reuter's Inc., the Fairlawn Plaza Style Center, Paul Breese, Ray Miller, the Cowabunga Safaris Square Table Gang, as well as various Cowabunga Safaris Alumni. And M. B.

Sincere appreciation to all my friends in Africa for their inspiration and goodwill.

Accolades to my family for their solidarity with a husband/father/grandpa who always seems to be working on the next book. I love you all.

The growth on this frog's keister is COWABUNGA, aka Gary K. Clarke, President-for-Life of Cowabunga Safaris.

Photo by Gary H. Lee